The Original AREA MAZES

100 Addictive Puzzles to Solve with Simple Math—and Clever Logic!

NAOKI INABA and RYOICHI MURAKAMI

Introduced by ALEX BELLOS

THE EXPERIMENT
NEW YORK

Originally published in Japan as *Menseki Meiro* (面積迷路) by Gakken Publishing Co., Ltd, Tokyo in 2011.
First published in North America by The Experiment, LLC, in 2017. English translation rights arranged with Gakken Plus Co., Ltd., through Paper Crane Agency.

The Experiment, LLC
220 East 23rd Street, Suite 301
New York, NY 10010-4674
theexperimentpublishing.com

ISBN 978-1-61519-421-6
Ebook ISBN 978-1-61519-422-3

Cover design by Lidija Tomas
Text design by Sarah Schneider

Manufactured in the United States of America
Distributed by Workman Publishing Company, Inc.
Distributed simultaneously in Canada by Thomas Allen & Son Ltd.

First printing October 2017
10 9 8 7 6 5 4 3 2 1

INTRODUCTION

NAOKI INABA, who was born in Nagoya in 1979, is one of the greatest puzzle inventors of all time, and he is not yet forty years old. He has already devised more types of pencil-and-paper logic puzzle than anyone else alive, and books of his area mazes have sold tens of thousands of copies in Japan.

Inaba's puzzle career began when he was a teenager. In the late '80s and '90s, two rival Japanese puzzle magazines–*Puzzler* and *Nikoli*–both encouraged readers to send in their own ideas. They inspired a generation of young puzzle designers, of whom Inaba became the most prolific. He was sixteen when he had his first invention published, and soon after that set up his own website where, for about ten years, he introduced a new puzzle type every week. Like sudoku and KenKen, these puzzles were all based on grids, to be filled in using logic alone. Each puzzle type had its own rules and required different strategies to solve.

Inaba is remarkable not only for the quantity of puzzle types he has invented but also for their minimalism and elegance. The rules can often be summed up in a single line. When Ryoichi Murakami, the director of El Camino coaching school in Tokyo, asked Inaba to devise a puzzle based on the areas of rectangles, he came up with *menseki meiro,* or area mazes. The challenge barely needs stating: Find the value of the question mark.

For a puzzle to catch on, however, it must not only have simple rules but also assume no technical expertise. Area mazes fulfill this requirement. All you need to know is that the area of a rectangle is the height times the width. No other mathematical techniques are needed–indeed, all others must be relinquished, especially fractions and equations.

This prohibition makes for a gorgeous and thrilling puzzle. Yes, we could plug the given values into a set of equations and solve them. But this strategy is laborious and ugly, plus it turns the puzzle into one that can be solved by rote. Instead, area mazes force us to be nimble and find the answer in the most elegant way. Solving them requires a burst of insight, not the same old tricks.

The area maze is one of only a few puzzles by Inaba that is not based on a grid, and the freedom to create examples of any shape adds an interesting unpredictability to the genre. With a sudoku, for example, there are no surprises: You might find it easy or difficult, but the path is clear. With an area maze, however, you are often surprised by the ingenuity of the solution. The feeling of satisfaction once you solve it is doubly pleasurable.

A good puzzle needs to come in all levels from easy to fiendish. Again, area mazes fit the bill. This book starts with elementary examples and ends with some that may have you scratching your head for days. Naoki Inaba has invented the perfect puzzle. Enjoy!

–ALEX BELLOS

ALEX BELLOS holds a degree in mathematics and philosophy from Oxford University. His bestselling books include *Here's Looking at Euclid* and *Can You Solve My Problems?* He is the coauthor of two coloring books, *Patterns of the Universe* and *Visions of the Universe*. He writes a popular math blog and a puzzle blog for the *Guardian*.

BRAIN TRAINING WITH AREA MAZES

As long as I keep my brain active, does it matter how?

What comes to mind when you hear the word "training"? Most likely a "workout" in which you use machines and barbells to strengthen your muscles. But today, more and more of us are "brain training"–stimulating the brain so that its function will not decline. It's a daily step you can take for your health–just like brushing your teeth.

Everything you do uses your brain somehow, so you might think it is enough to just focus your mind on any task. However, some activities are better than others.

Puzzles with math and logic in mind

El Camino is a science and math coaching school for students from first grade through high school. To prepare our elementary school students for the Mathematical Olympiads, we give them puzzles that strengthen their logical and creative thinking.

Saying "let's solve a puzzle!" appeals to the children much more than "let's do a math problem." The students tackle the puzzles enthusiastically–having fun, developing their abilities, and deepening their interest in mathematics all at once.

We invented "area mazes" (*menseki meiro*) especially for our third-grade students. Area mazes can seem impossible to solve without using fractions and decimals. However, our third-grade students haven't learned those techniques yet, so of course you can solve them using only whole numbers! The challenge is to work out *how*. It takes more than calculation: It takes logic, spatial reasoning, and wits.

Getting the answer *without* doing any complicated math is what makes area mazes fun!

Isn't it too late for my brain to grow?

Let me return to the original topic. When our students' parents ask me how to train their own brains, I recommend the puzzles we use at El Camino. You might say, "What? But those are for kids!" Yes, that is true–and that's exactly my point.

The young mind is a flexible "blank slate," whereas the adult brain becomes inflexible. That's why many puzzle books designed for adults fall short: The puzzles tend to follow a pattern. Once your brain gets used to it, you will focus on spotting and applying the pattern–rather than thinking creatively.

To use our metaphor of strength training, doing that kind of puzzle is something like rehab. Repeating an action over and over again will keep your brain active and maintain its function to some extent. However, it cannot rejuvenate your brain: It may put off aging, but it won't make you younger.

Area mazes are different: They are designed to develop flexible thinking–youthful thinking, if you will. They cannot be solved by repeating a process. You often need a stroke of inspiration to solve them!

As you work through these puzzles, you will feel your brain "waking up." I hope you will enjoy area mazes. There is a delightful sense of achievement that comes with saying, "I got it!"

–**RYOICHI MURAKAMI,** director of El Camino

HOW TO SOLVE AREA MAZES

Using the given lengths and areas, find the value of ?. Remember, the formula for the area of a rectangle is height × width.

If your calculation creates a fraction or decimal, STOP and look for another way. Area mazes can be solved using whole numbers only!*

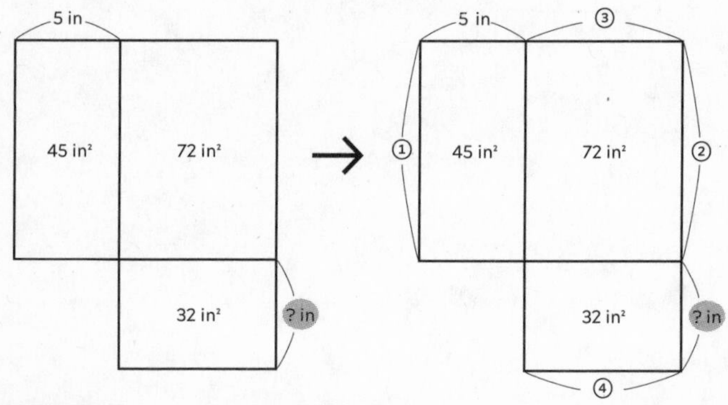

EXAMPLE ONE

Find length ① . . . 45 ÷ 5 = 9 in.

Find length ② . . . This is the same as ①, so 9 in.

Find length ③ . . . 72 ÷ 9 = 8 in.

Find length ④ . . . This is the same as ③, so 8 in.

Length ? is 32 ÷ 8 = 4 in.

SOLUTION: 4 in.

*However, do not assume that *every* length or area in the puzzle must be a whole number.

EXAMPLE TWO

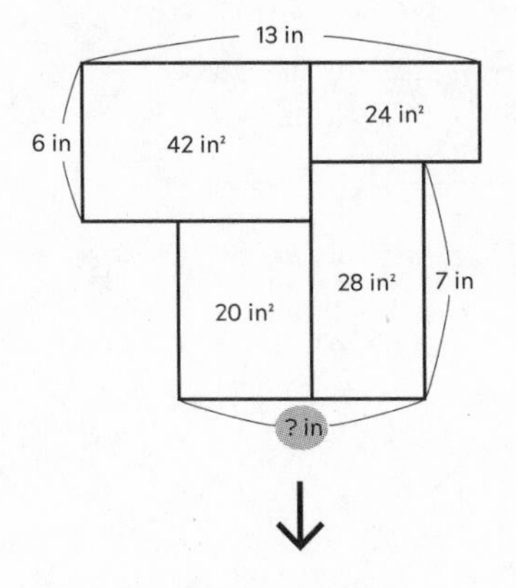

Note that the figures are not drawn to scale. You can't solve by "eyeballing"—you have to prove it with math!

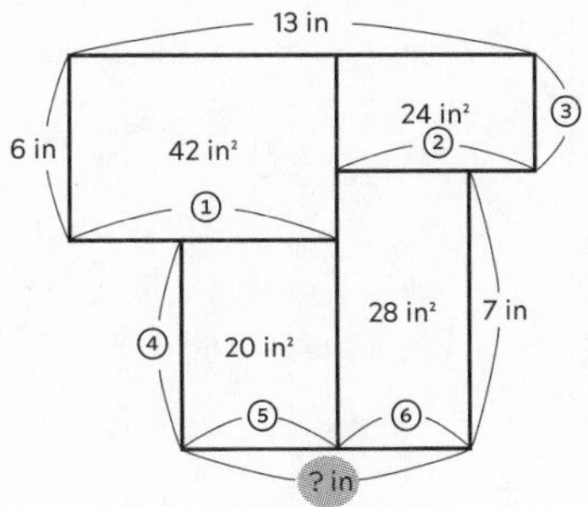

Even after you have solved a problem, you can revisit it to look for a more elegant solution.

Find length ① . . . 42 ÷ 6 = 7 in.

Find length ② . . . 13 – 7 = 6 in.

Find length ③ . . . 24 ÷ 6 = 4 in.

Find length ④ . . . (4 + 7) – 6 = 5 in.

Find length ⑤ . . . 20 ÷ 5 = 4 in.

Find length ⑥ . . . 28 ÷ 7 = 4 in.

Length ? is 4 + 4 = 8 in.

SOLUTION: 8 in.

EXAMPLE THREE

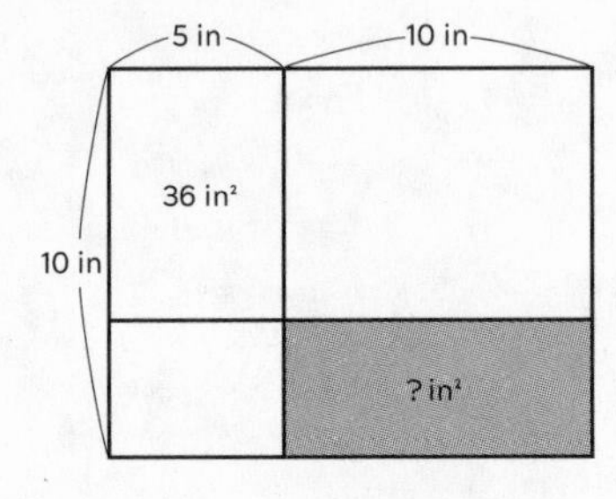

First, consider the two rectangles on the left together.

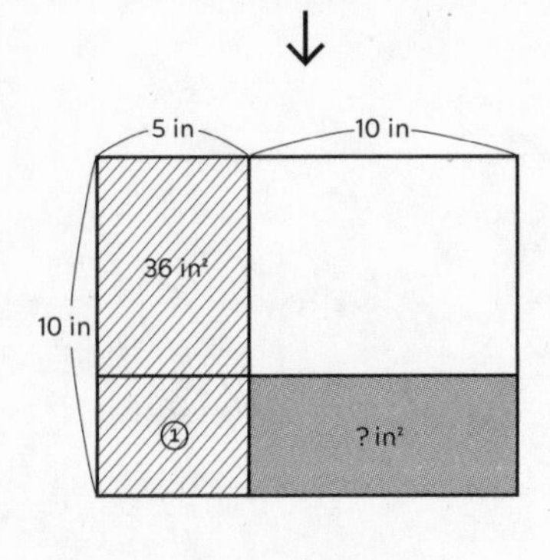

Find the area with stripes . . . $5 \times 10 = 50$ in.^{2}

Find area ① . . . $50 - 36 = 14$ in.^{2}

Next, we will look at the two rectangles on the bottom.

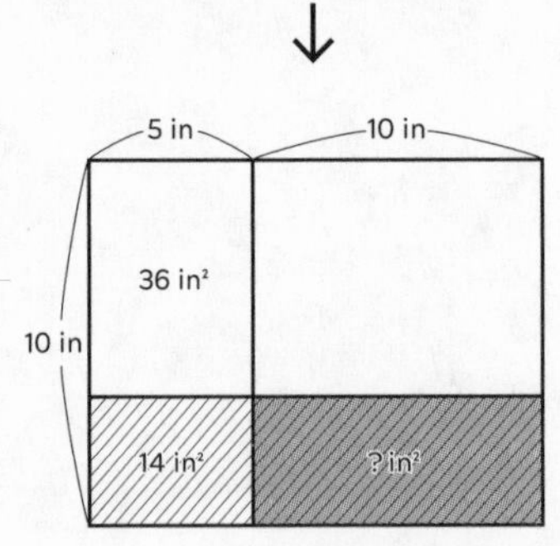

Area ? is the same height as area ①, and exactly twice as wide.

So, area ? must be exactly double area ①.

Area ? is $14 \times 2 = 28$ in.^{2}

SOLUTION: 28 in.^{2}

PUZZLES

The puzzles get more challenging as you go.
This key will guide you!

LEVEL 1

LEVEL 2

LEVEL 3

LEVEL 4

LEVEL 5

PUZZLE 1

Find the solution on page 112.

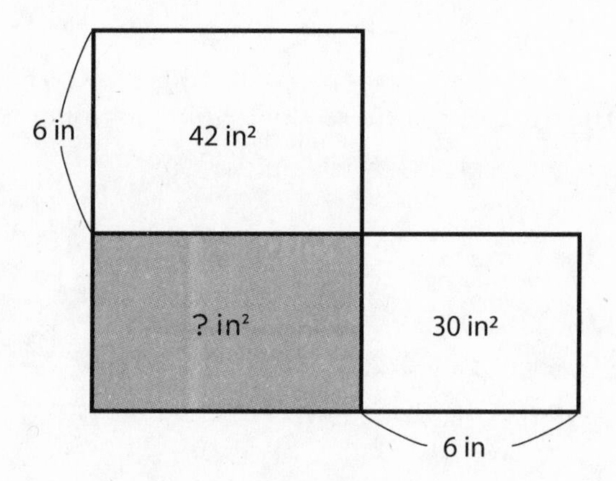

SOLUTION

PUZZLE 2

Find the solution on page 112.

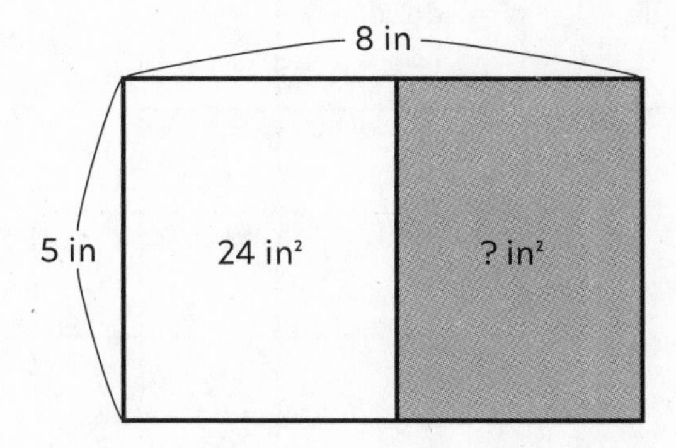

SOLUTION

PUZZLE 3

Find the solution on page 112.

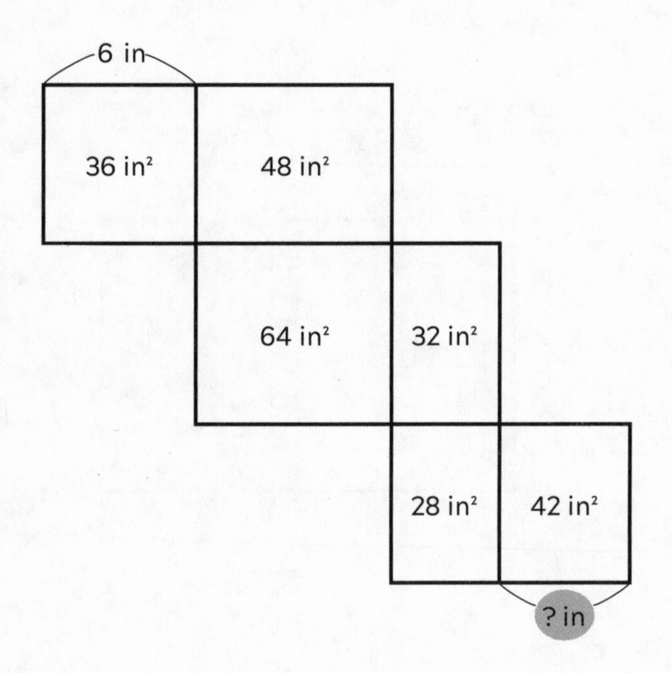

SOLUTION

PUZZLE 4

Find the solution on page 113.

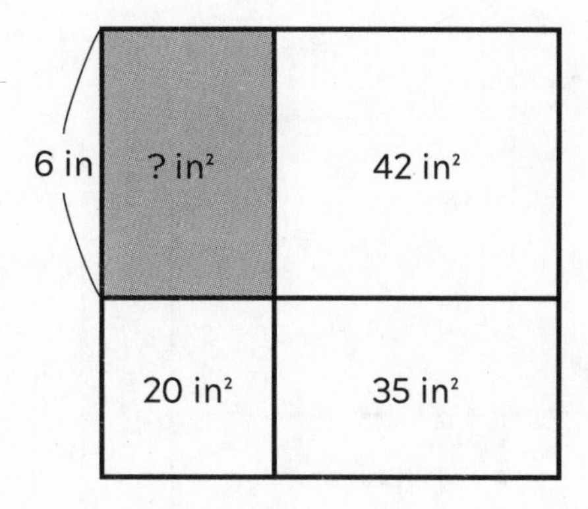

SOLUTION

PUZZLE 5

Find the solution on page 113.

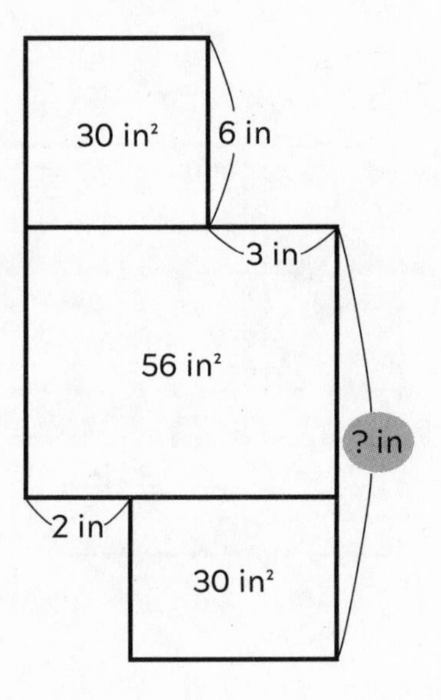

SOLUTION

____

PUZZLE 6

Find the solution on page 113.

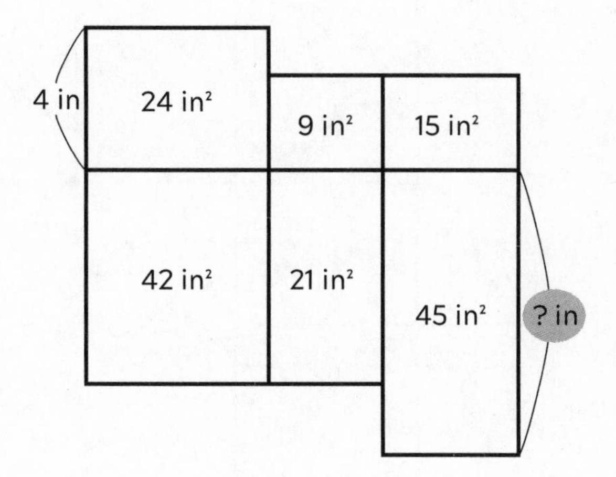

SOLUTION

PUZZLE 7

Find the solution on page 114.

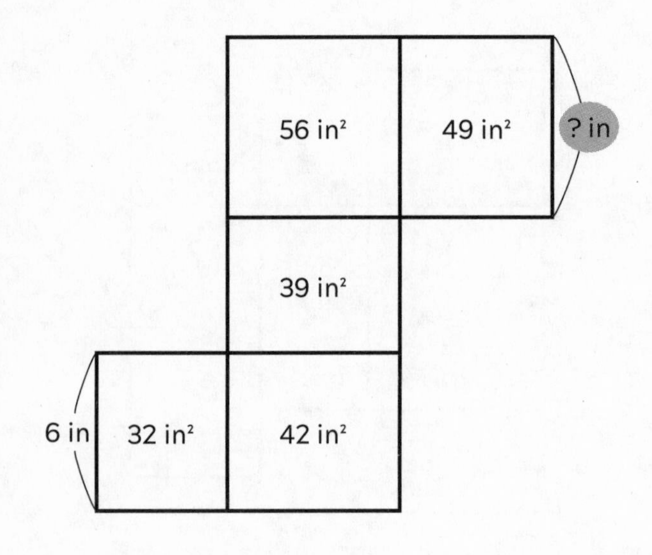

SOLUTION

PUZZLE 8

Find the solution on page 114.

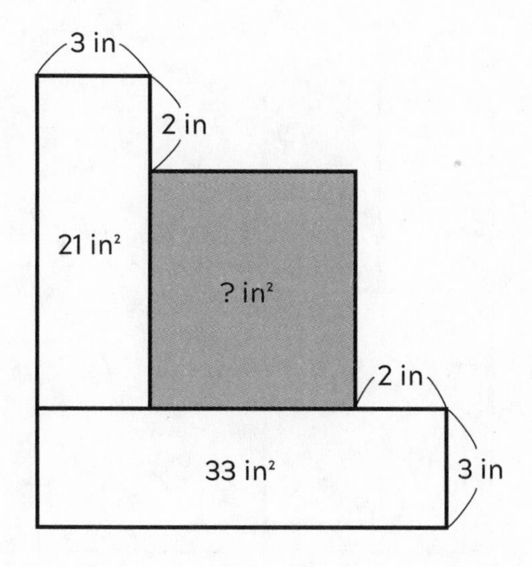

SOLUTION

PUZZLE 9

Find the solution on page 114.

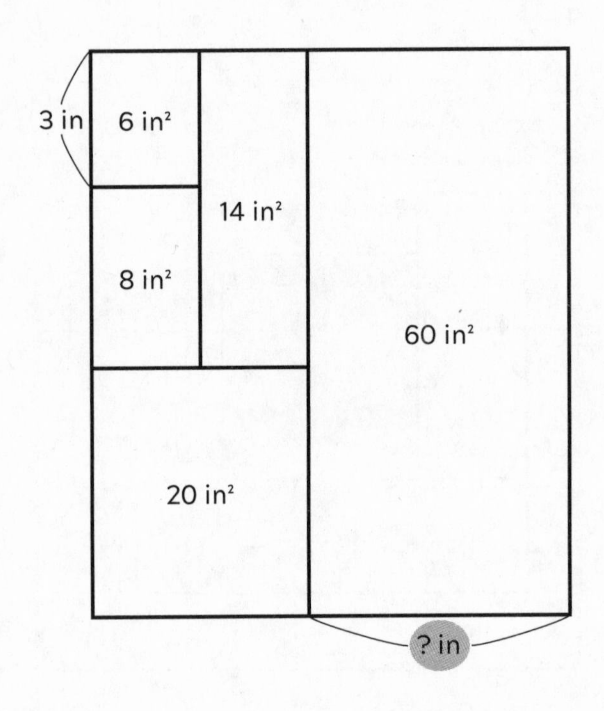

SOLUTION

PUZZLE 10

Find the solution on page 115.

5 in

20 in^2		16 in^2	
	? in^2	24 in^2	
35 in^2			42 in^2
	32 in^2		48 in^2

SOLUTION

PUZZLE 11

Find the solution on page 115.

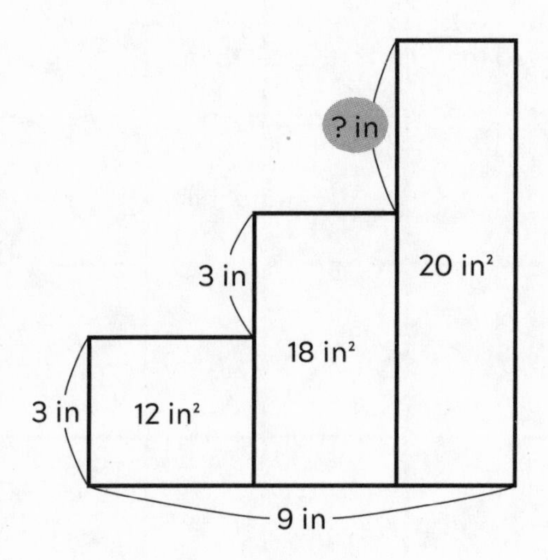

SOLUTION

PUZZLE 12

Find the solution on page 115.

6 in: ? in²	48 in²	
30 in²		20 in²
	72 in²	36 in²

SOLUTION

PUZZLE 13

Find the solution on page 116.

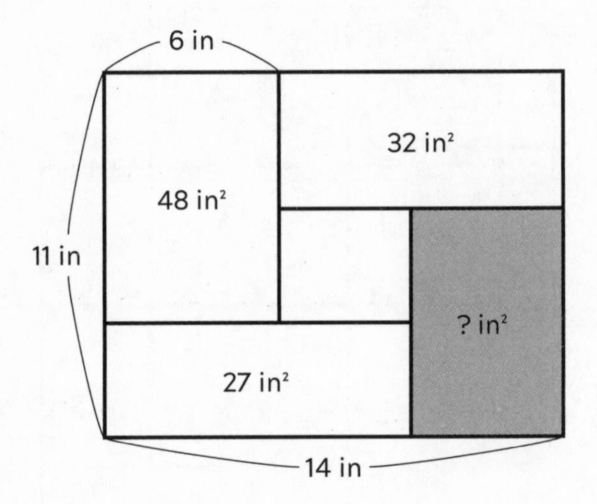

SOLUTION

PUZZLE 14

Find the solution on page 116.

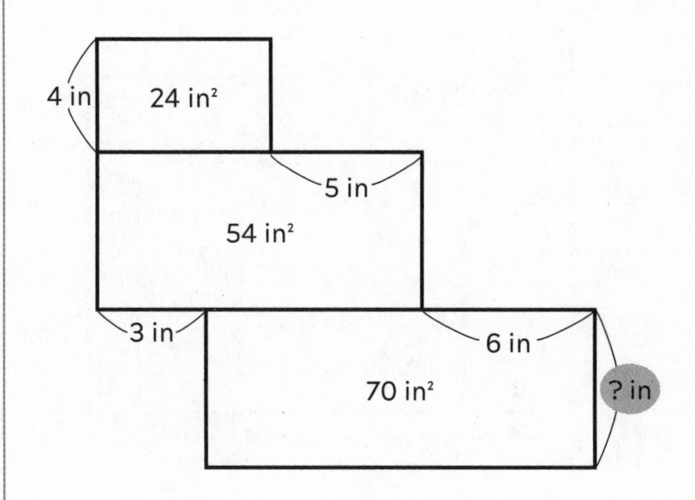

SOLUTION

PUZZLE 15

Find the solution on page 116.

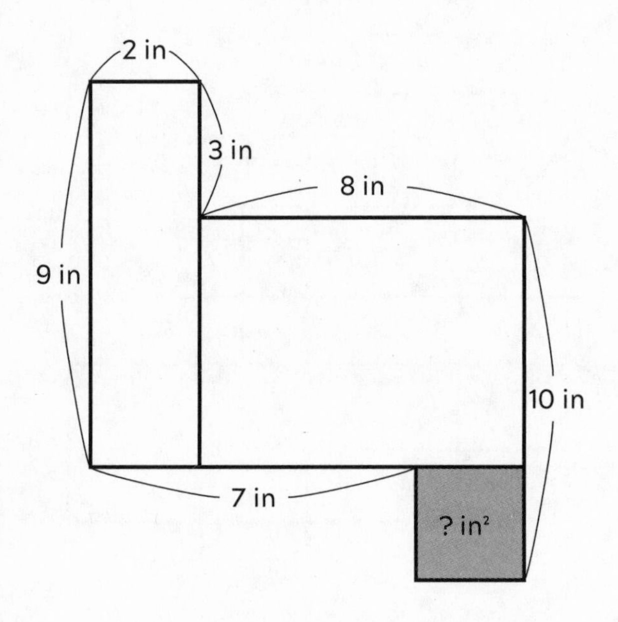

SOLUTION

PUZZLE 16

Find the solution on page 117.

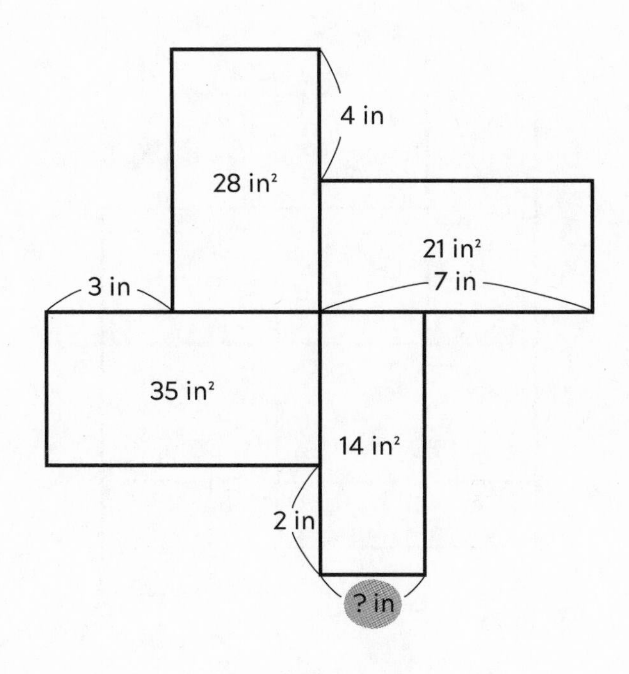

SOLUTION

PUZZLE 17

Find the solution on page 117.

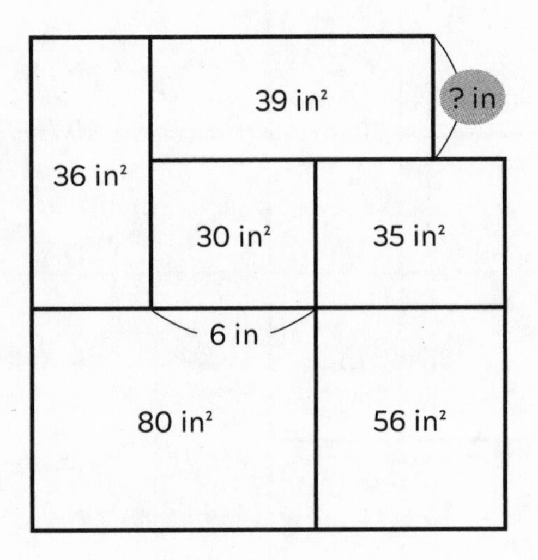

SOLUTION

PUZZLE 18

Find the solution on page 117.

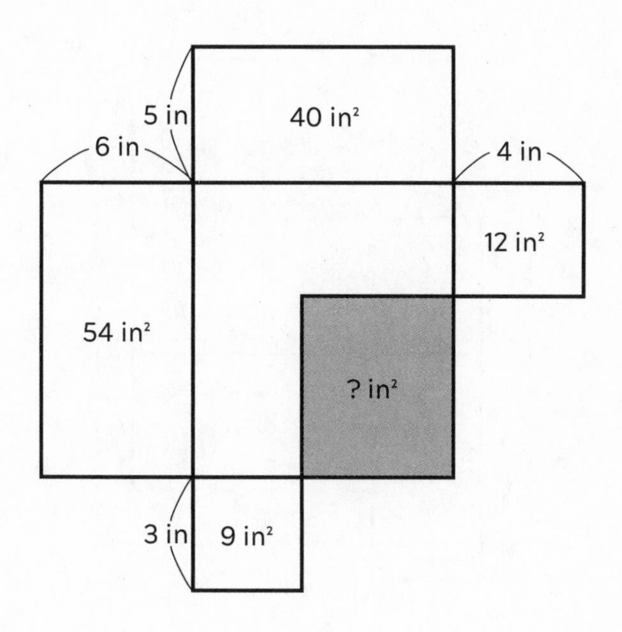

SOLUTION

PUZZLE 19

Find the solution on page 118.

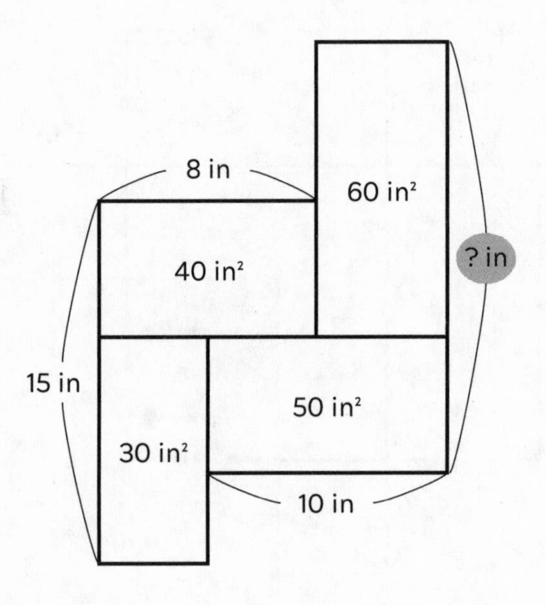

SOLUTION

PUZZLE 20

Find the solution on page 118.

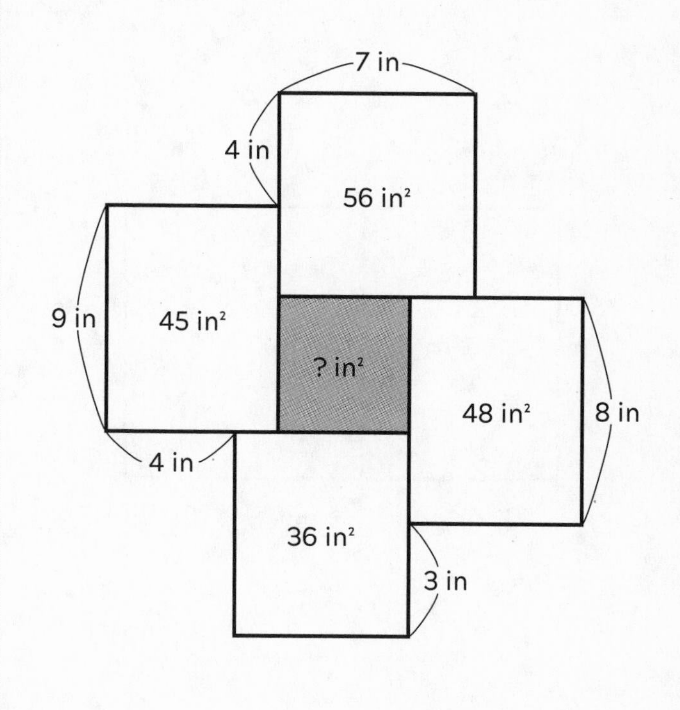

SOLUTION

PUZZLE 21

Find the solution on page 118.

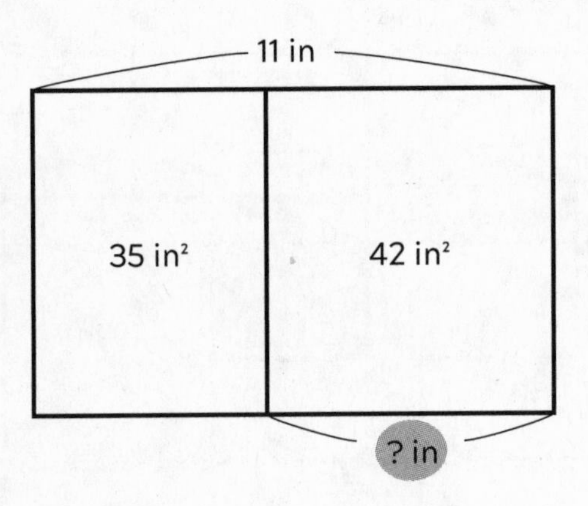

SOLUTION

____

PUZZLE 22

Find the solution on page 119.

13 in²	26 in²
15 in²	? in²

SOLUTION

____

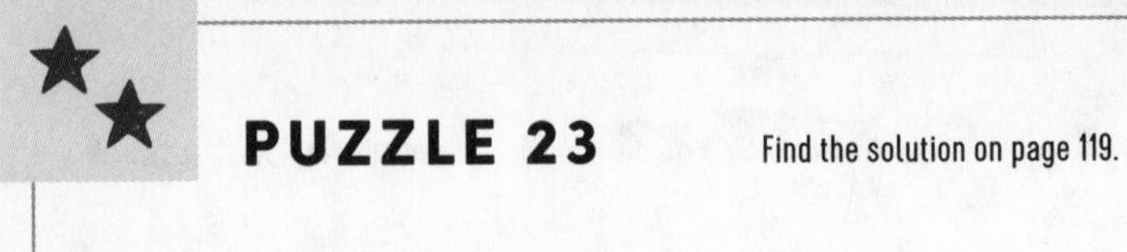

PUZZLE 23

Find the solution on page 119.

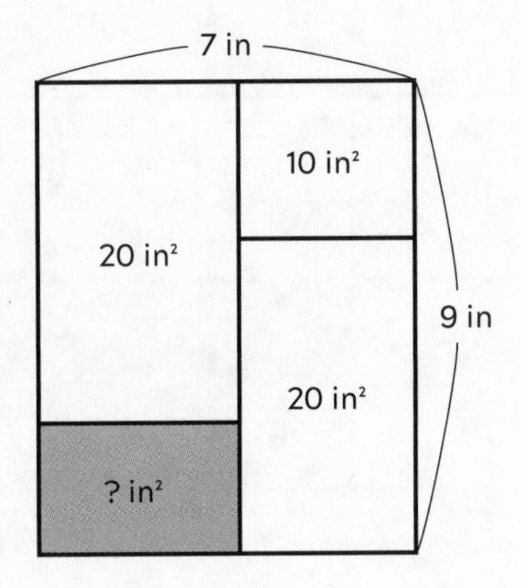

SOLUTION

____

PUZZLE 24

Find the solution on page 119.

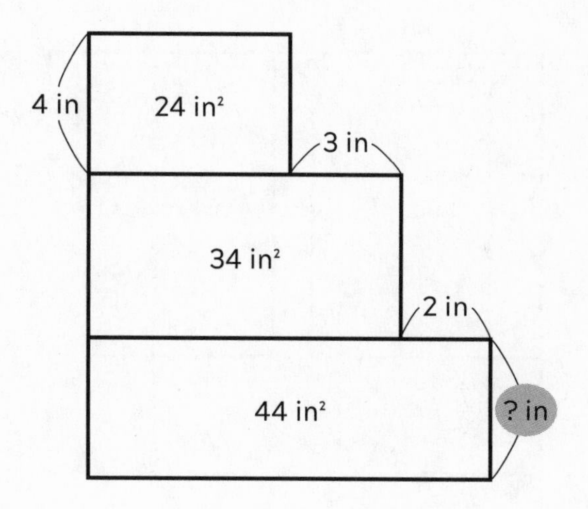

SOLUTION

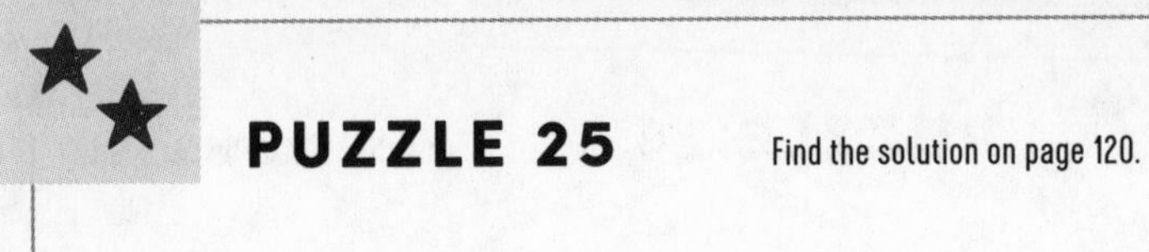

PUZZLE 25

Find the solution on page 120.

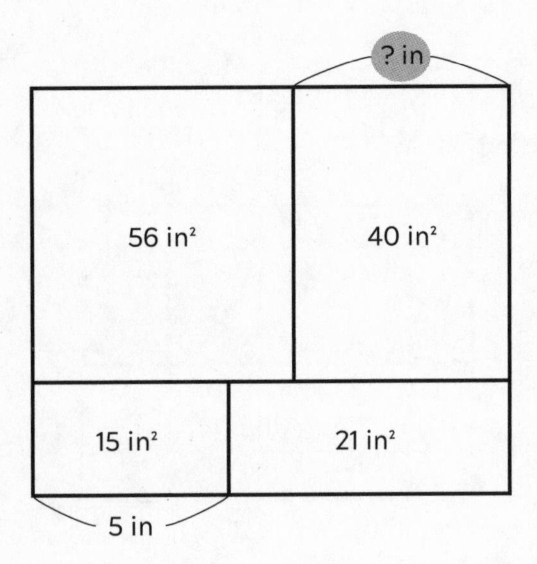

SOLUTION

PUZZLE 26

Find the solution on page 120.

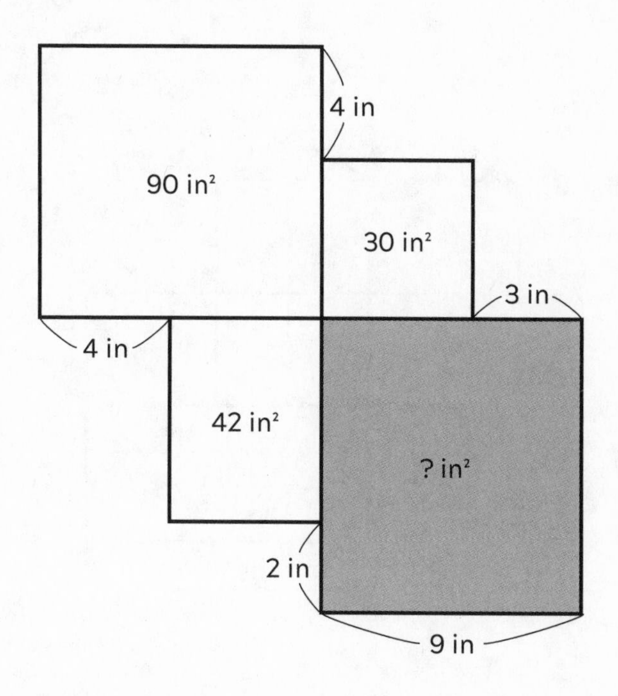

SOLUTION

PUZZLE 27

Find the solution on page 120.

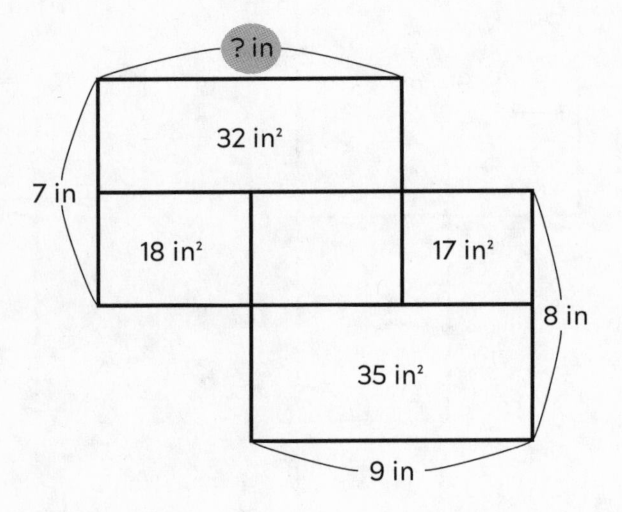

SOLUTION

PUZZLE 28

Find the solution on page 121.

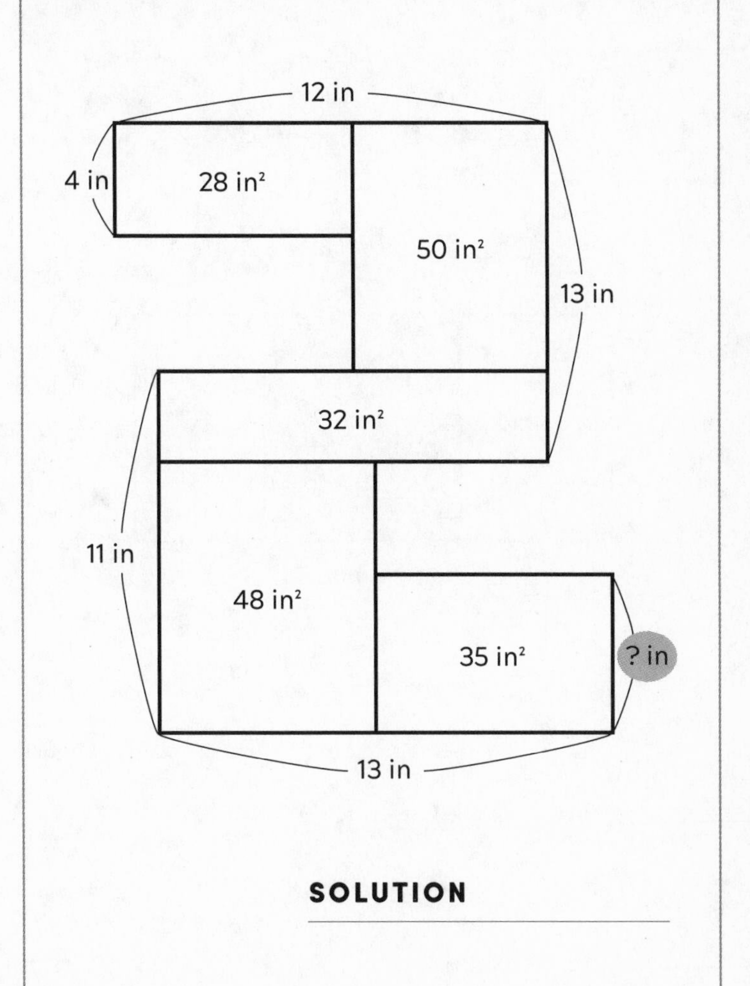

SOLUTION

PUZZLE 29

Find the solution on page 121.

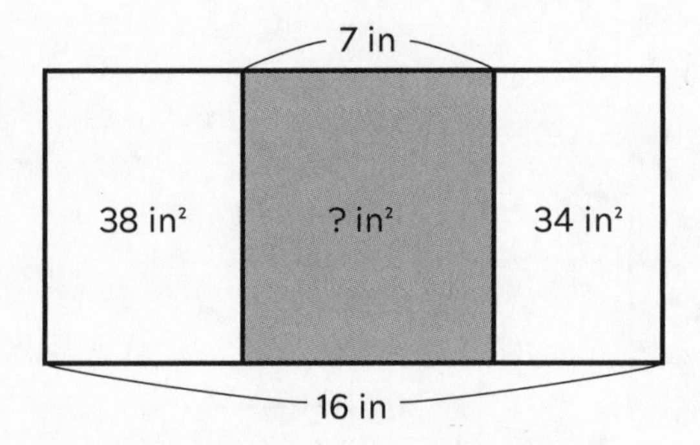

SOLUTION

PUZZLE 30

Find the solution on page 121.

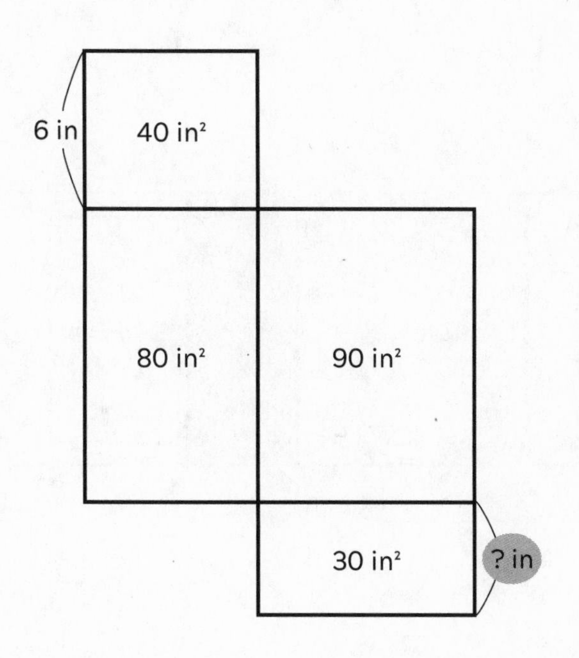

SOLUTION

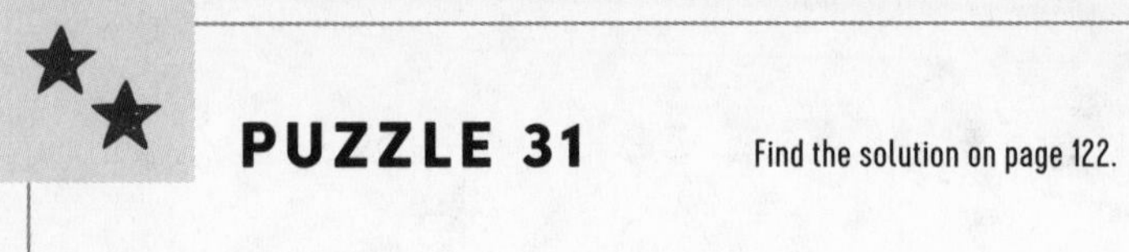

PUZZLE 31

Find the solution on page 122.

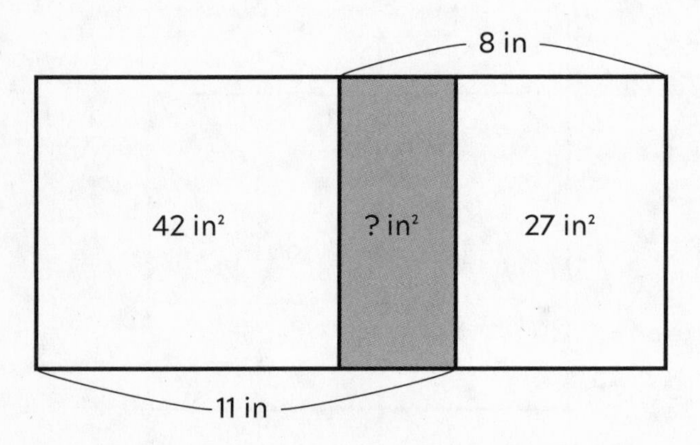

SOLUTION

PUZZLE 32

Find the solution on page 122.

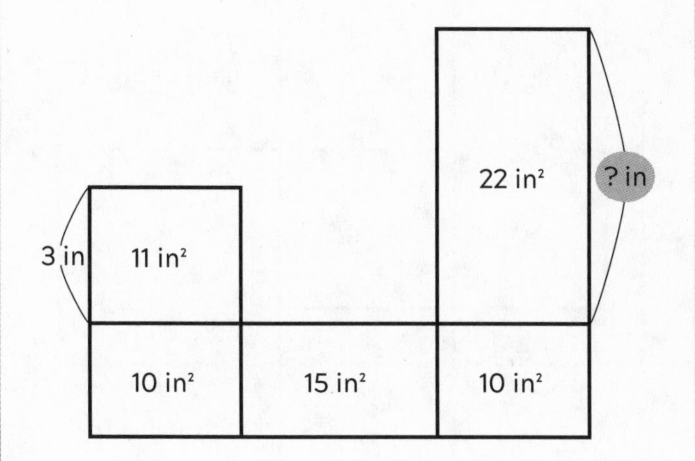

SOLUTION

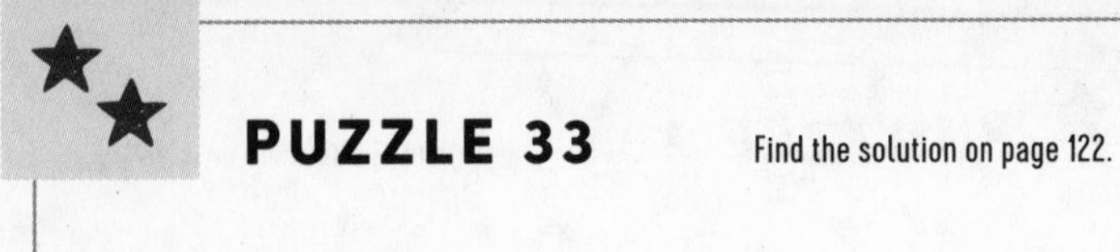

PUZZLE 33

Find the solution on page 122.

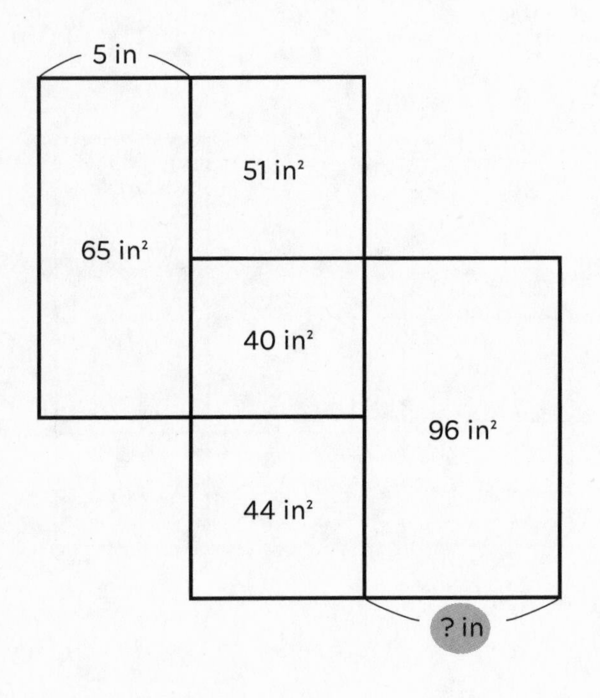

SOLUTION

PUZZLE 34

Find the solution on page 123.

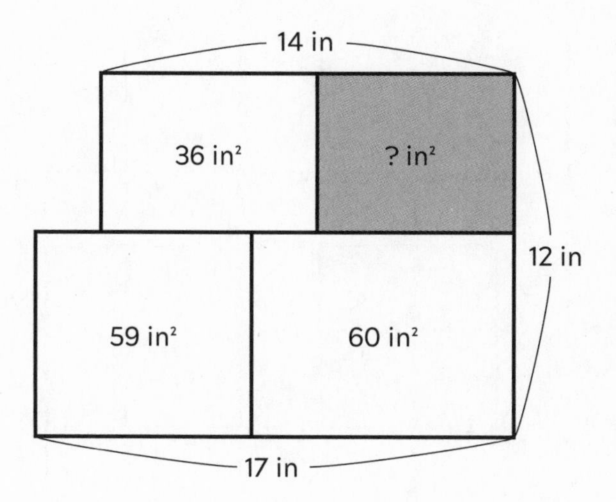

SOLUTION

PUZZLE 35

Find the solution on page 123.

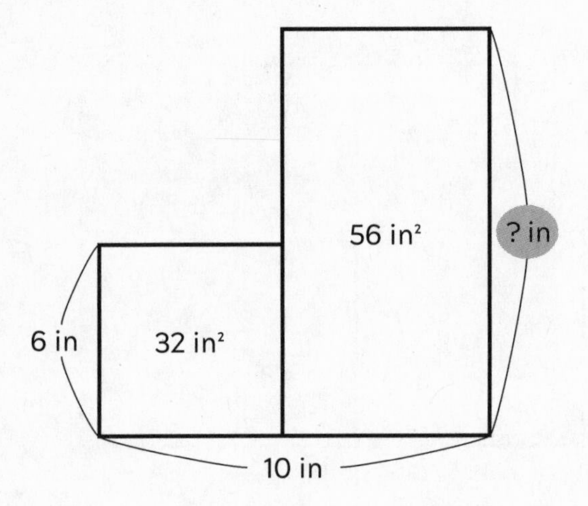

SOLUTION

PUZZLE 36

Find the solution on page 123.

6 in

28 in²	50 in²	24 in²
? in²	70 in²	28 in²

SOLUTION

PUZZLE 37

Find the solution on page 124.

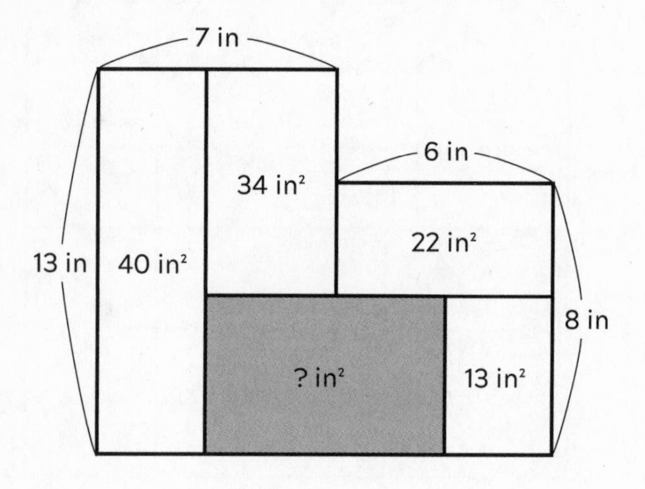

SOLUTION

PUZZLE 38

Find the solution on page 124.

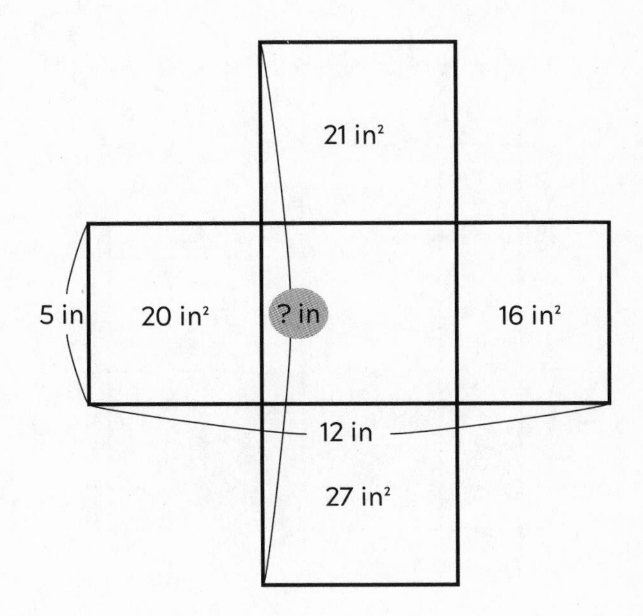

SOLUTION

PUZZLE 39

Find the solution on page 125.

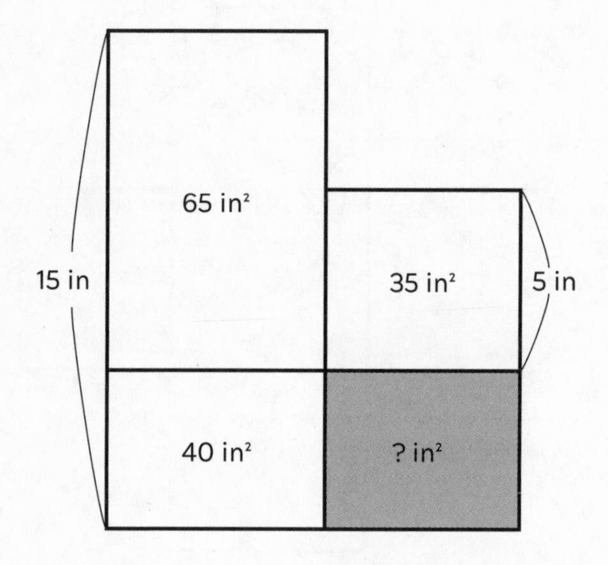

SOLUTION

PUZZLE 40

Find the solution on page 125.

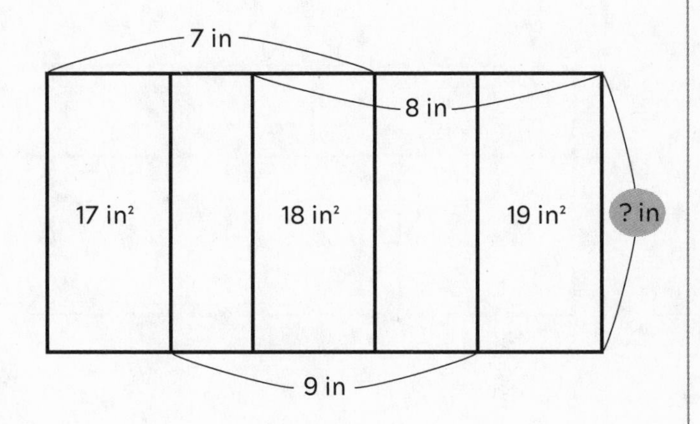

SOLUTION

PUZZLE 41

Find the solution on page 125.

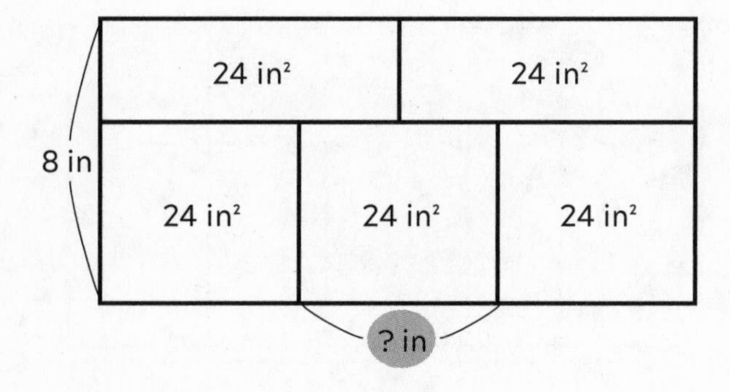

SOLUTION

PUZZLE 42

Find the solution on page 126.

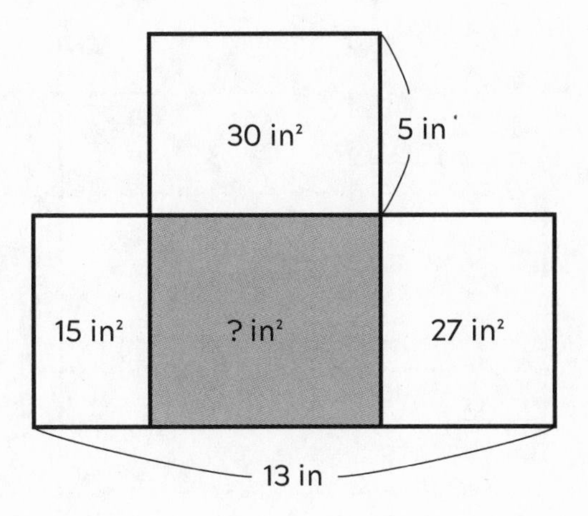

SOLUTION

PUZZLE 43

Find the solution on page 126.

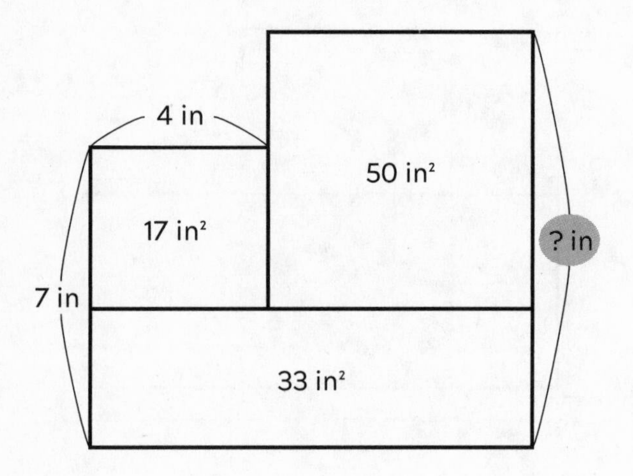

SOLUTION

PUZZLE 44

Find the solution on page 127.

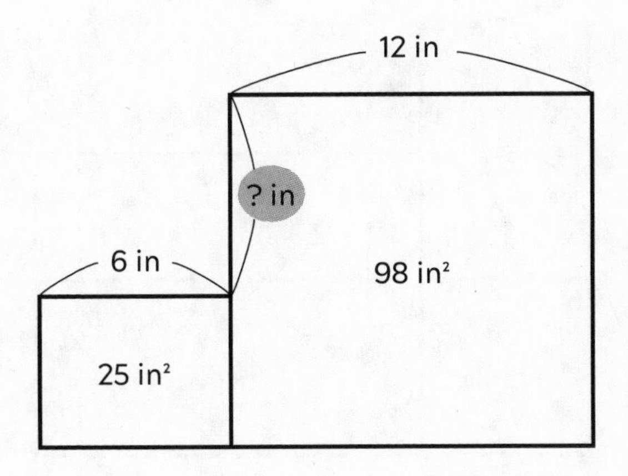

SOLUTION

PUZZLE 45

Find the solution on page 127.

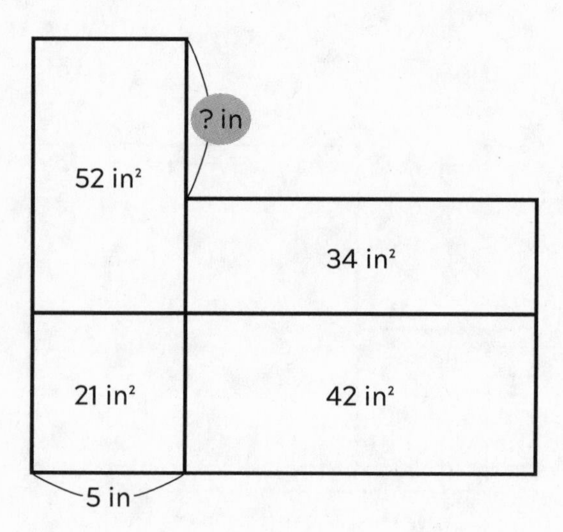

SOLUTION

PUZZLE 46

Find the solution on page 128.

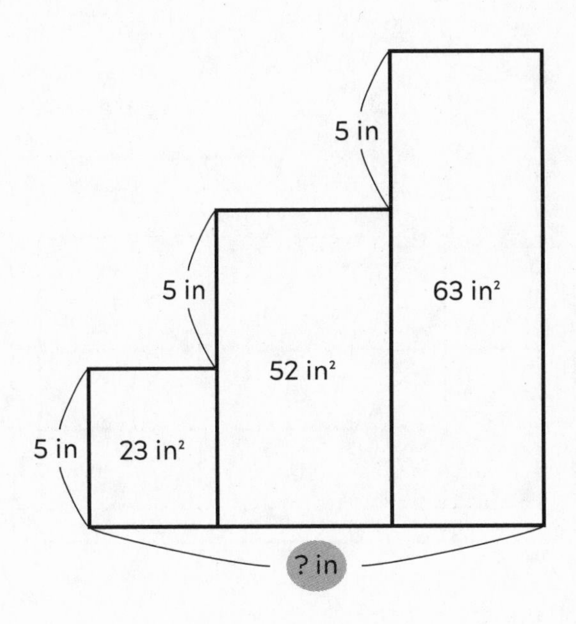

SOLUTION

PUZZLE 47

Find the solution on page 128.

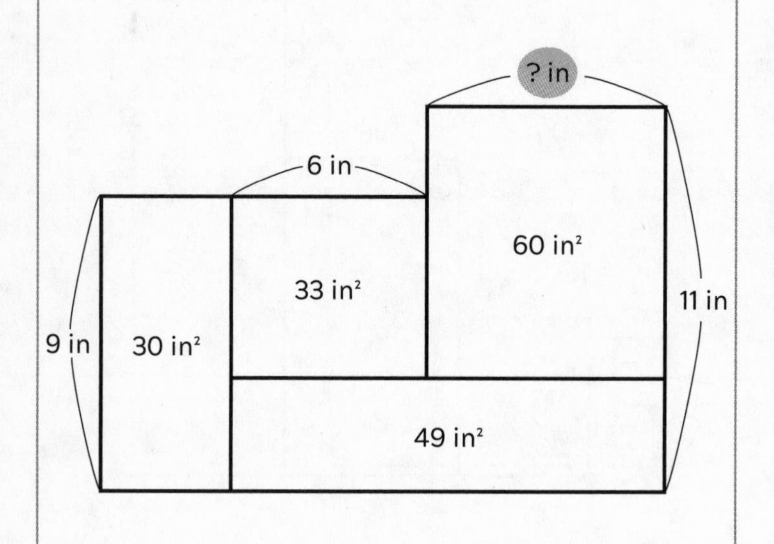

SOLUTION

PUZZLE 48

Find the solution on page 129.

13 in²

26 in²

38 in²

40 in²

? in²

10 in²

SOLUTION

PUZZLE 49

Find the solution on page 129.

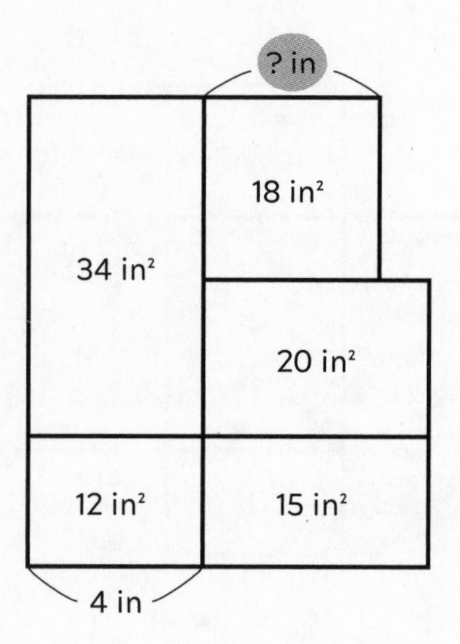

SOLUTION

____

PUZZLE 50

Find the solution on page 130.

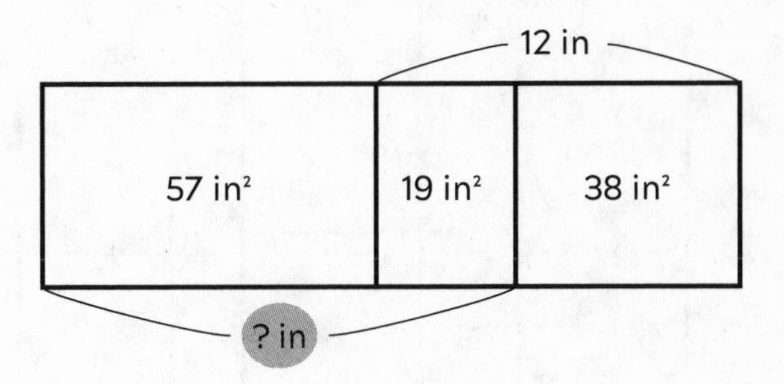

SOLUTION

PUZZLE 51

Find the solution on page 130.

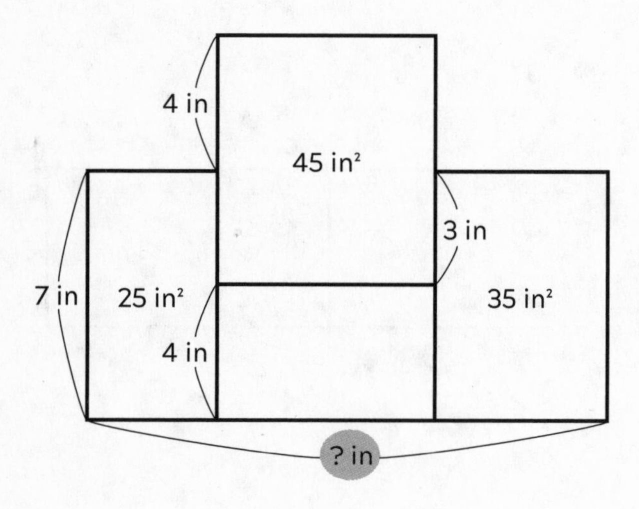

SOLUTION

PUZZLE 52

Find the solution on page 131.

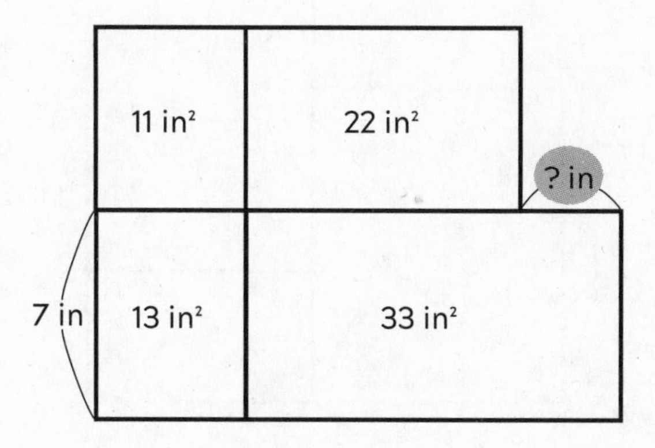

SOLUTION ____________________

PUZZLE 53

Find the solution on page 131.

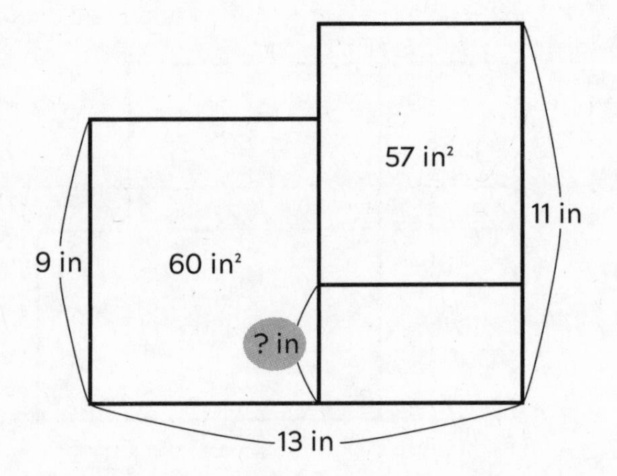

SOLUTION

PUZZLE 54

Find the solution on page 132.

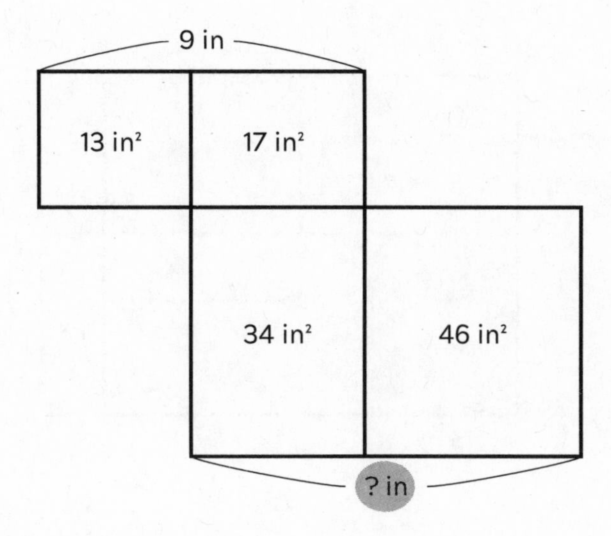

SOLUTION

PUZZLE 55

Find the solution on page 132.

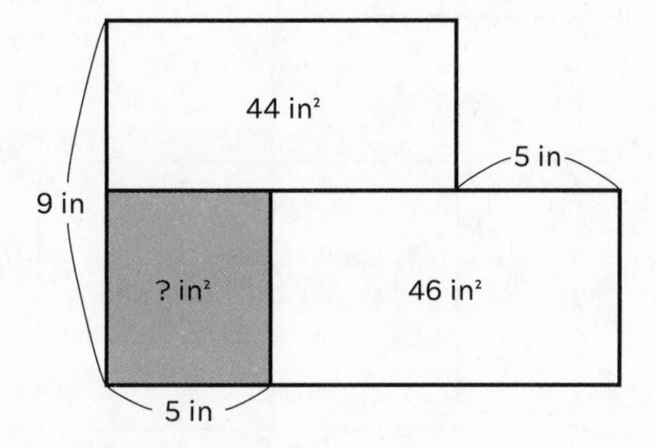

SOLUTION

PUZZLE 56

Find the solution on page 133.

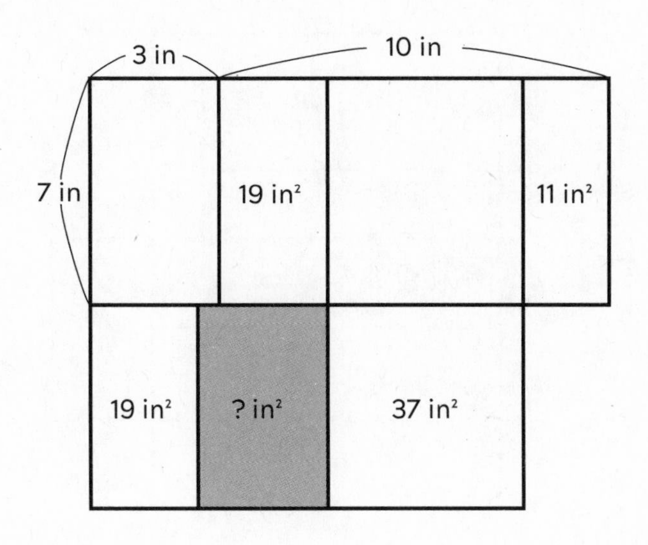

SOLUTION

PUZZLE 57

Find the solution on page 133.

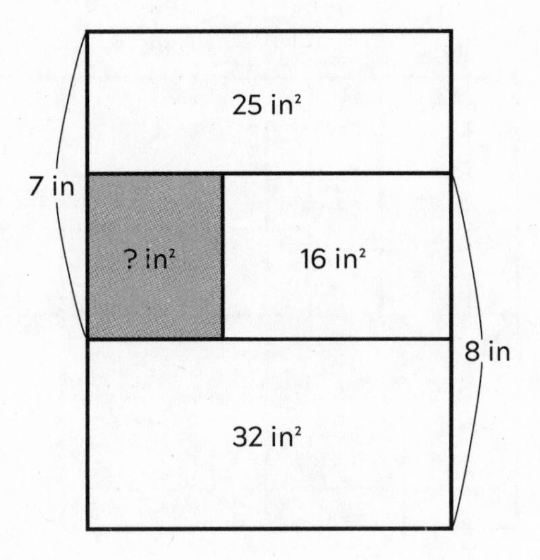

SOLUTION

PUZZLE 58

Find the solution on page 134.

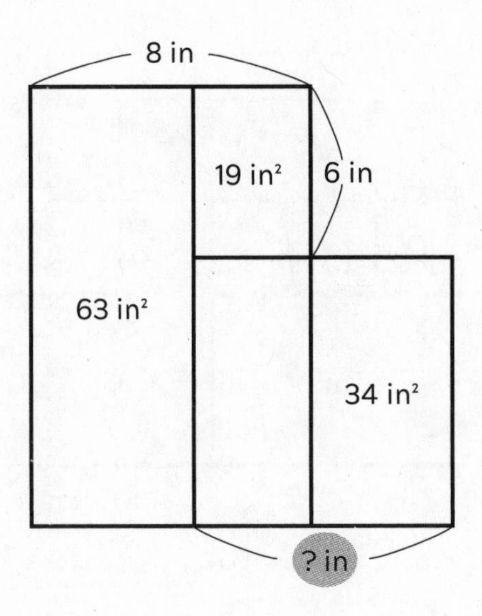

SOLUTION

PUZZLE 59

Find the solution on page 134.

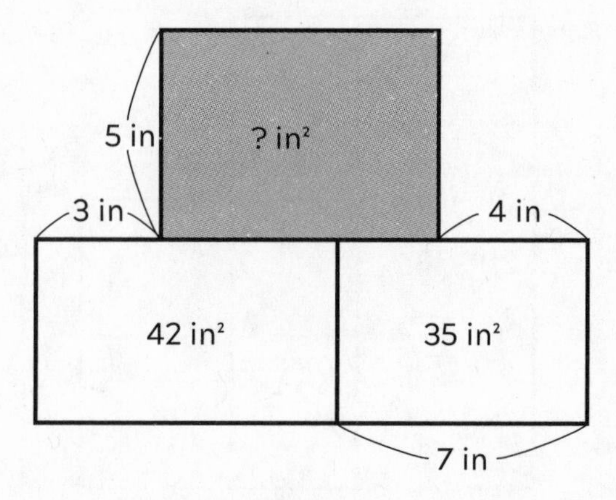

SOLUTION

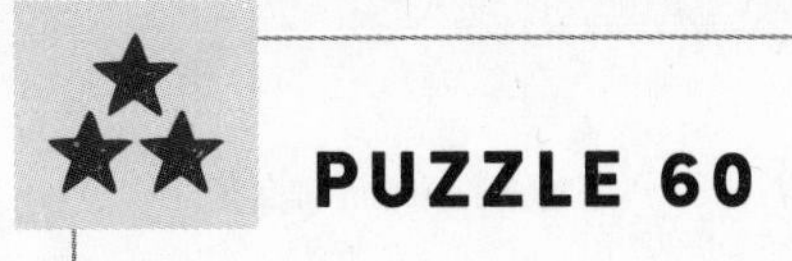

PUZZLE 60

Find the solution on page 135.

26 in² 17 in²

? in²

13 in²

34 in²

68 in²

SOLUTION

PUZZLE 61

Find the solution on page 135.

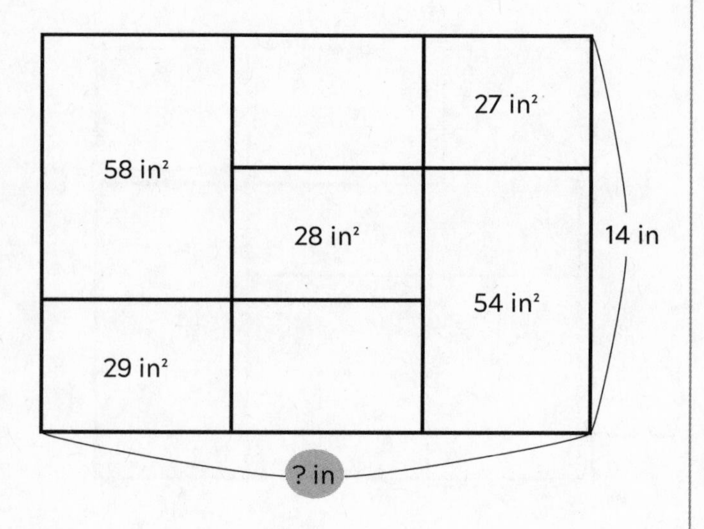

SOLUTION

PUZZLE 62

Find the solution on page 136.

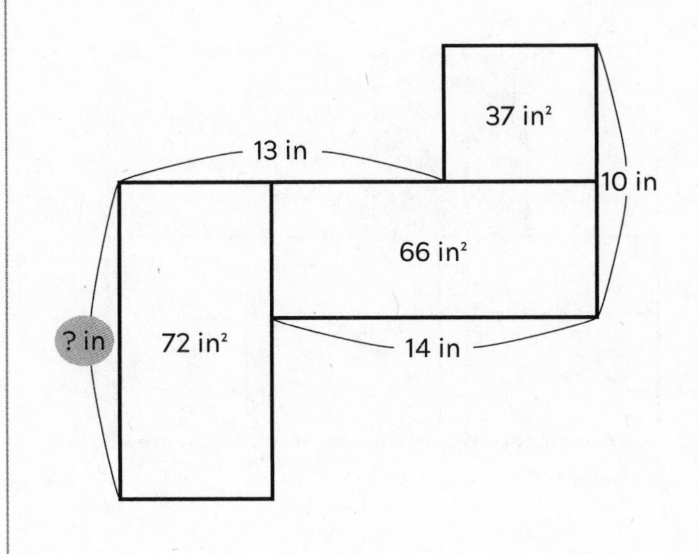

SOLUTION

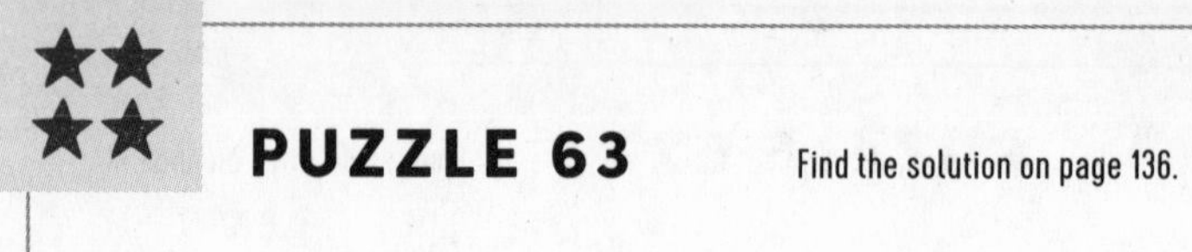

PUZZLE 63

Find the solution on page 136.

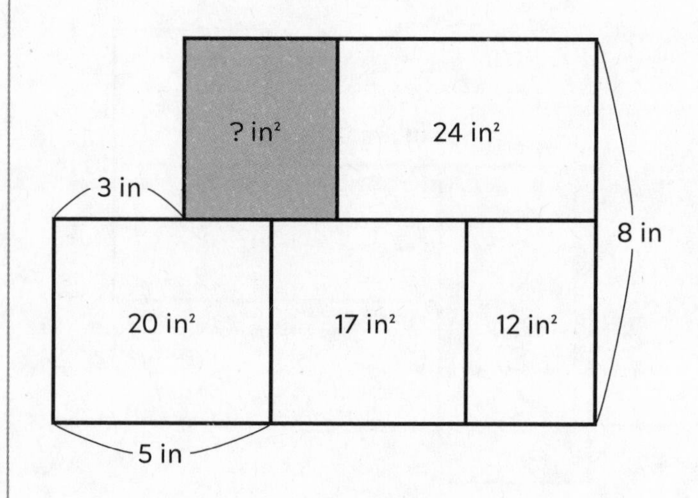

SOLUTION

PUZZLE 64

Find the solution on page 137.

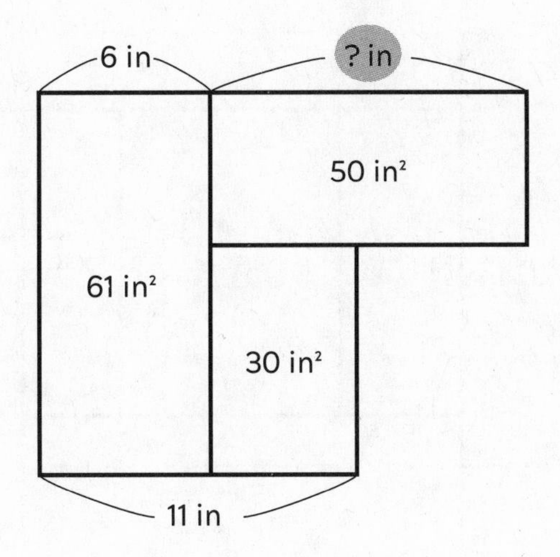

SOLUTION

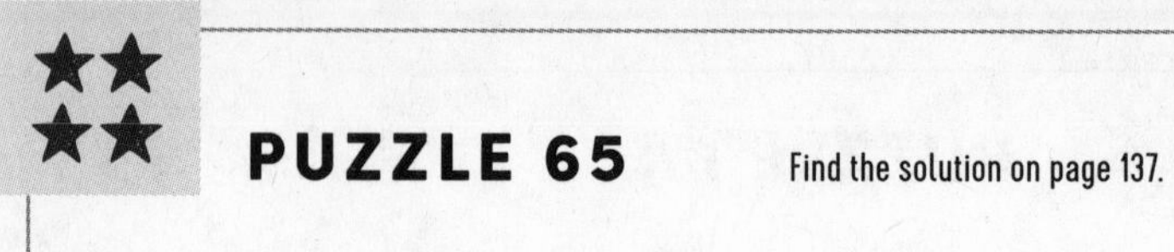

PUZZLE 65

Find the solution on page 137.

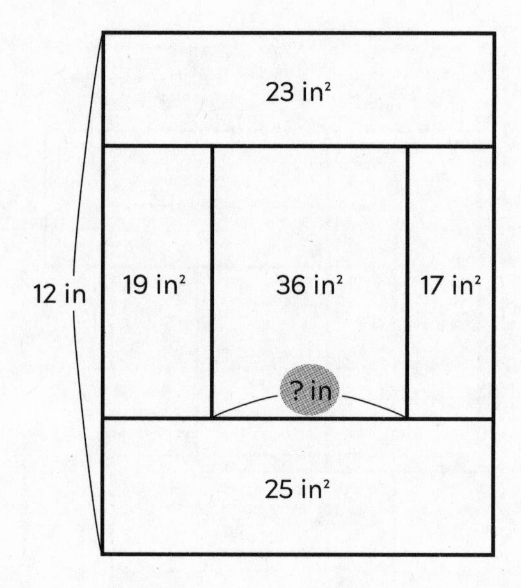

SOLUTION

PUZZLE 66

Find the solution on page 138.

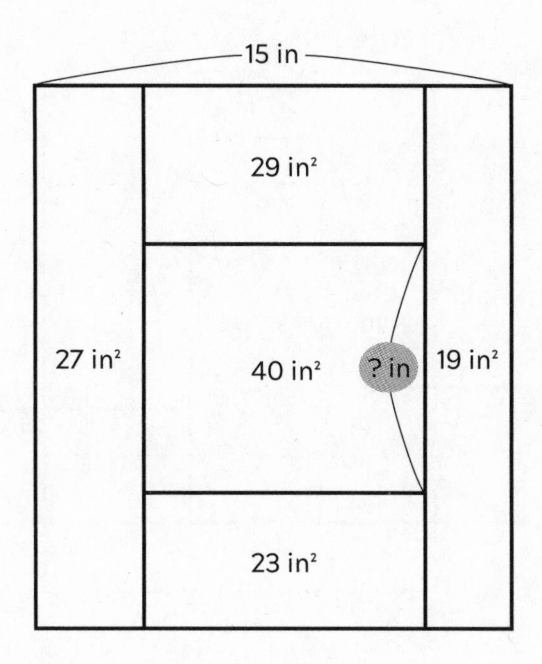

SOLUTION

PUZZLE 67

Find the solution on page 138.

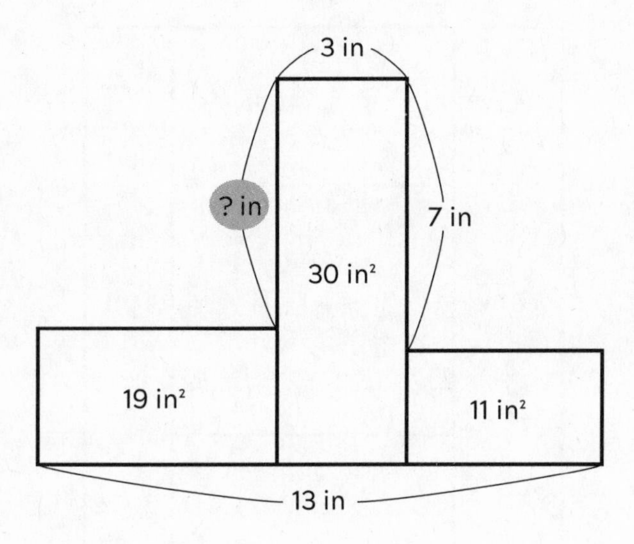

SOLUTION

PUZZLE 68

Find the solution on page 139.

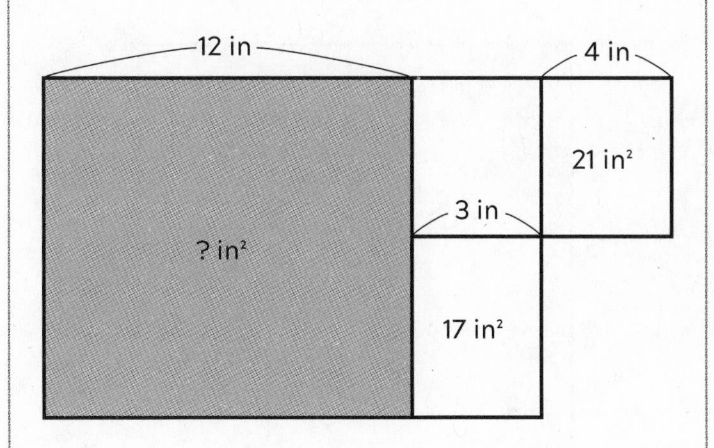

SOLUTION

____

PUZZLE 69

Find the solution on page 139.

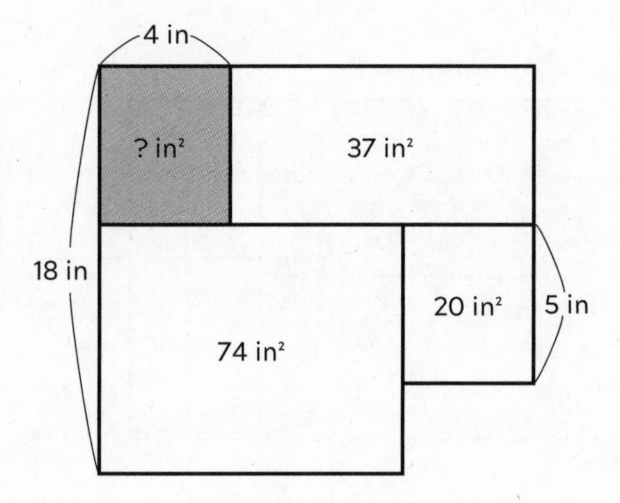

SOLUTION

PUZZLE 70

Find the solution on page 140.

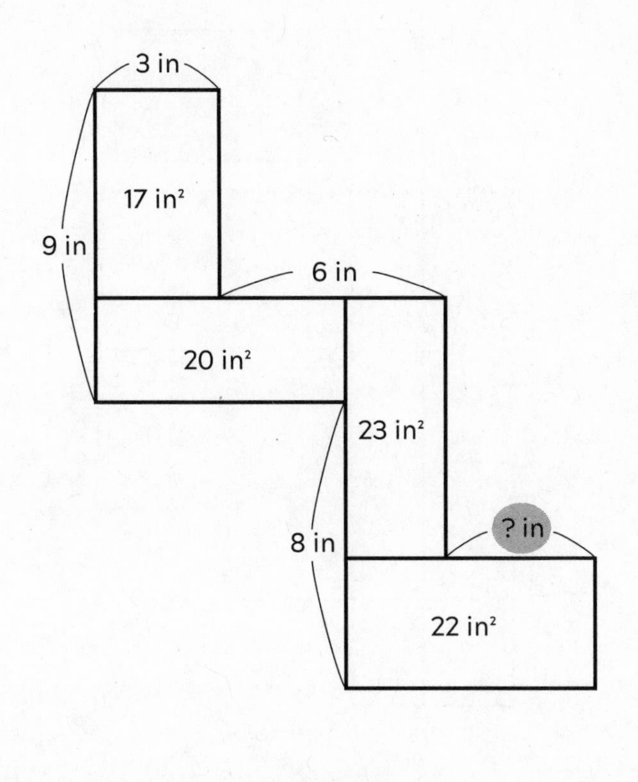

SOLUTION

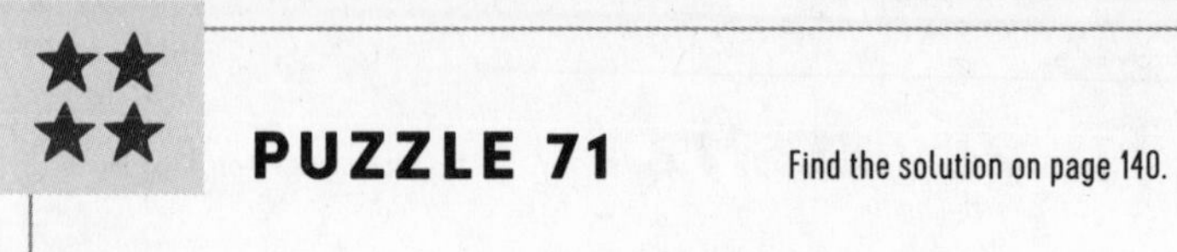

PUZZLE 71

Find the solution on page 140.

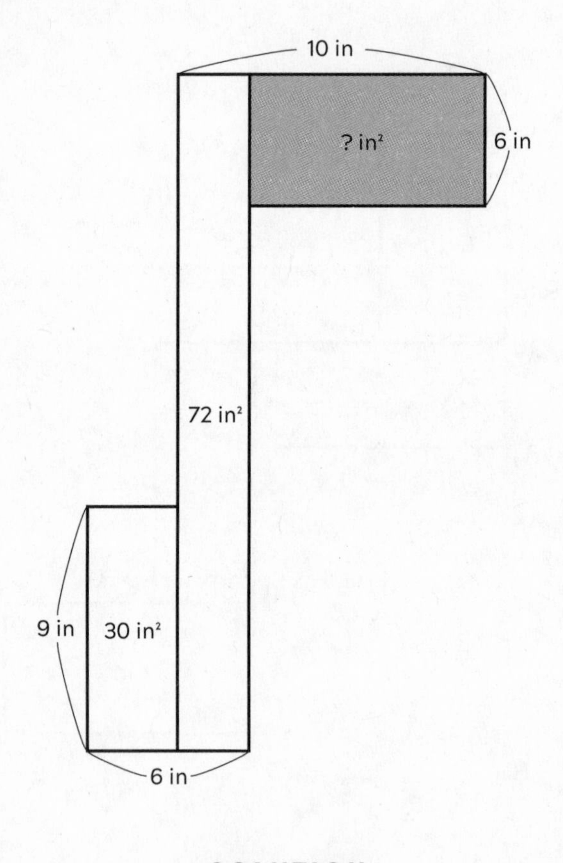

SOLUTION

PUZZLE 72

Find the solution on page 141.

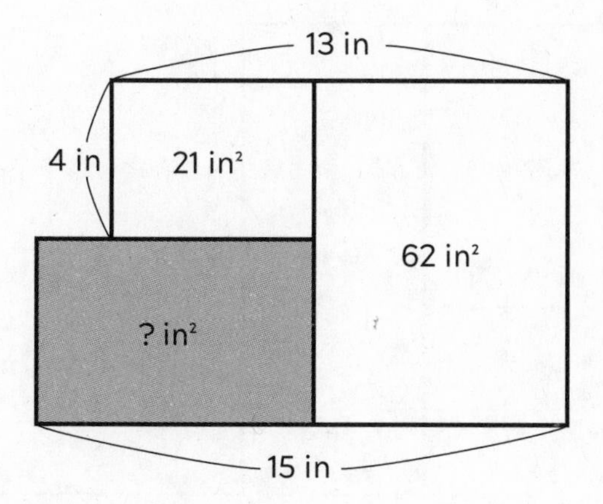

SOLUTION

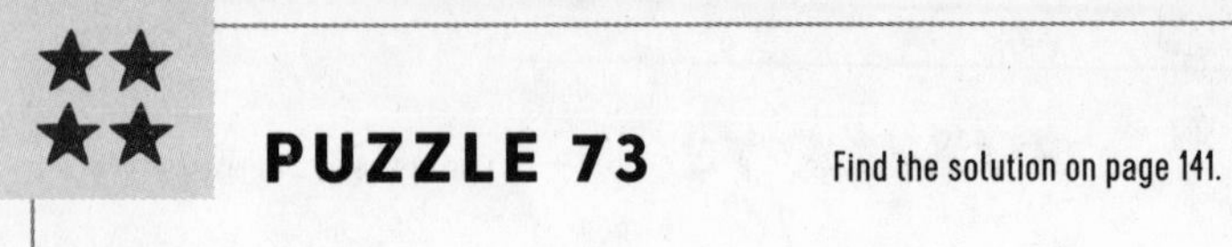

PUZZLE 73

Find the solution on page 141.

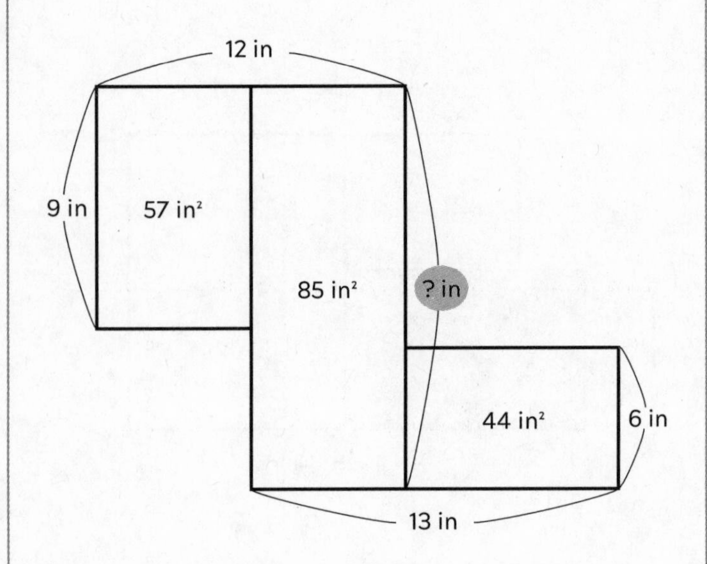

SOLUTION

PUZZLE 74

Find the solution on page 142.

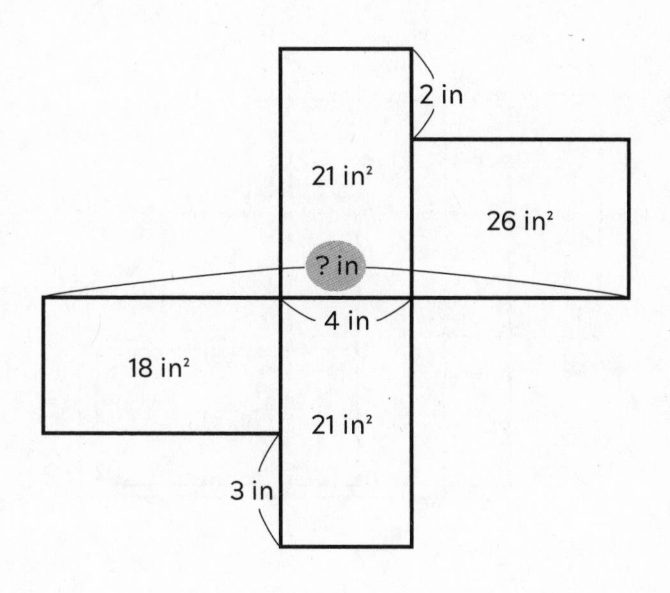

SOLUTION

PUZZLE 75

Find the solution on page 142.

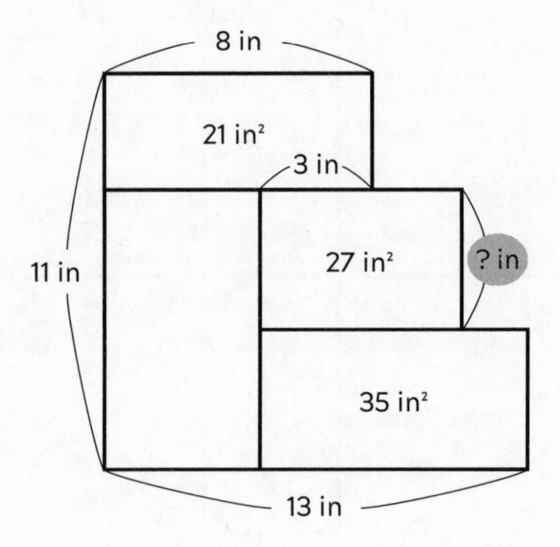

SOLUTION ______________________

PUZZLE 76

Find the solution on page 143.

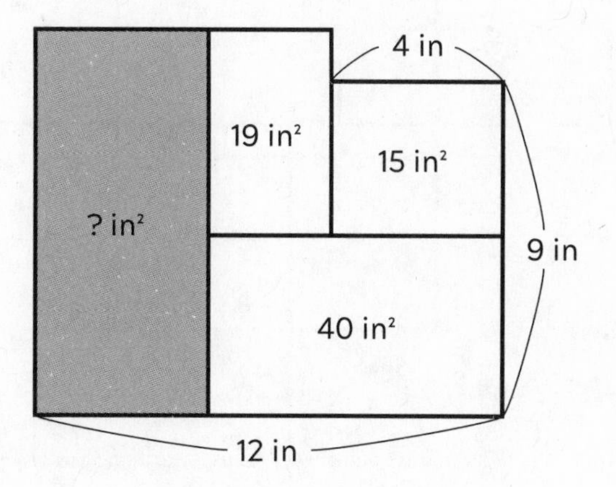

SOLUTION

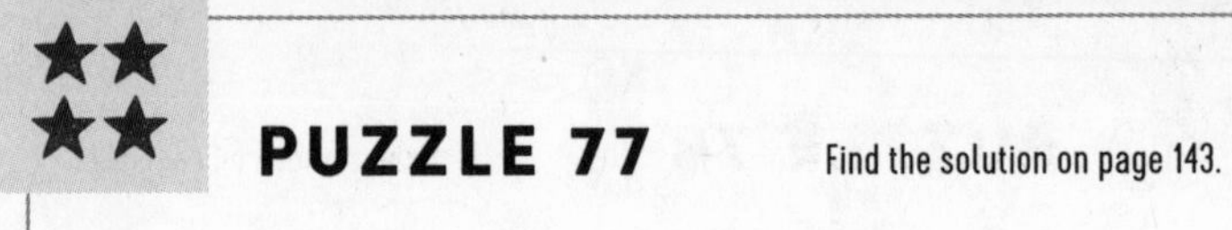

PUZZLE 77

Find the solution on page 143.

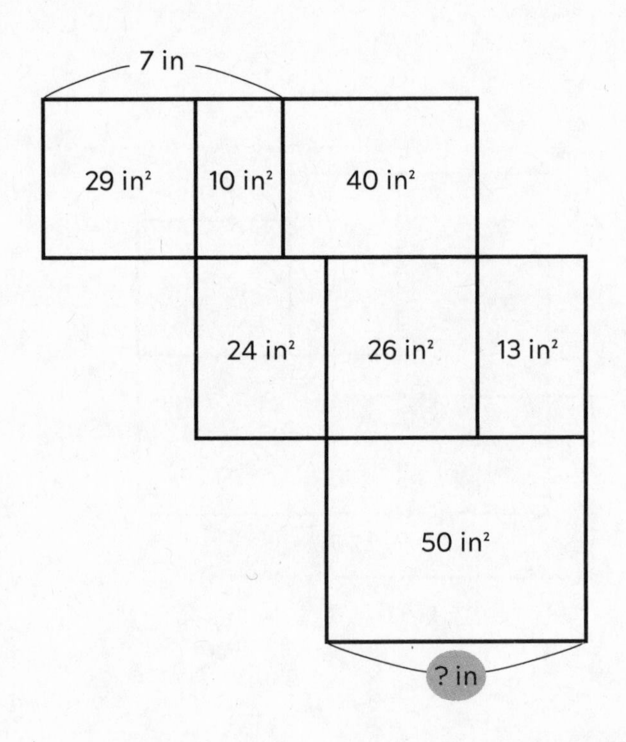

SOLUTION

PUZZLE 78

Find the solution on page 144.

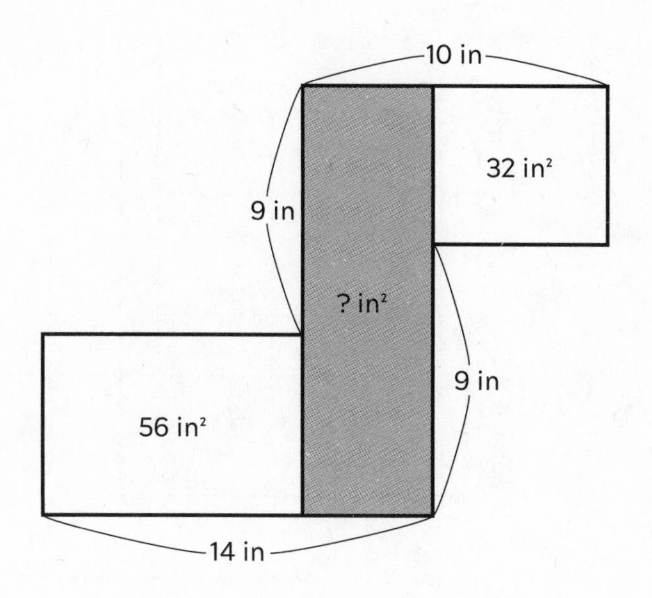

SOLUTION

PUZZLE 79

Find the solution on page 144.

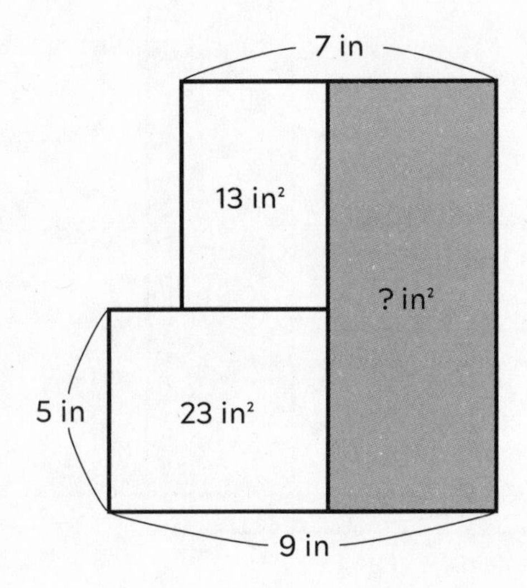

SOLUTION

PUZZLE 80

Find the solution on page 145.

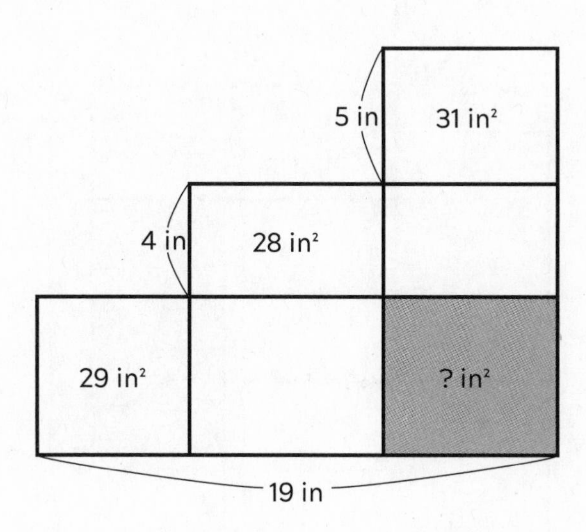

SOLUTION

PUZZLE 81

Find the solution on page 145.

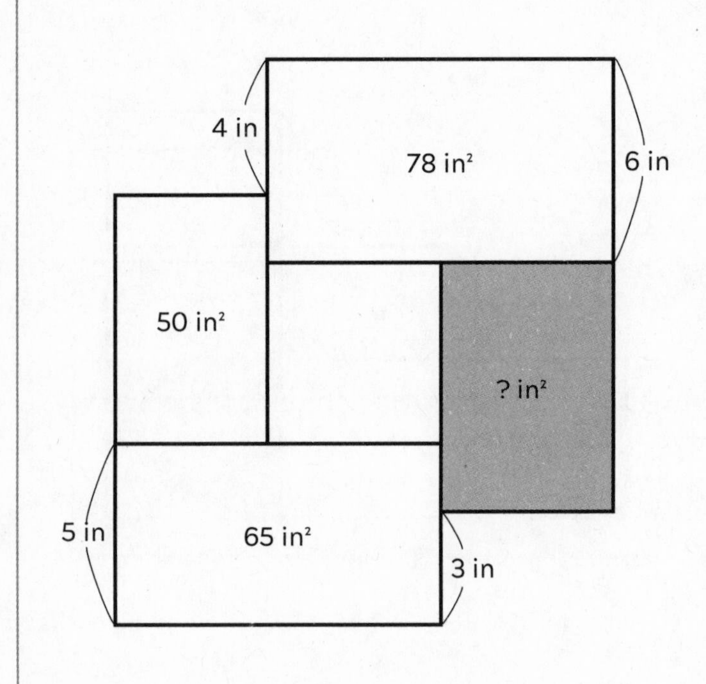

SOLUTION

PUZZLE 82

Find the solution on page 146.

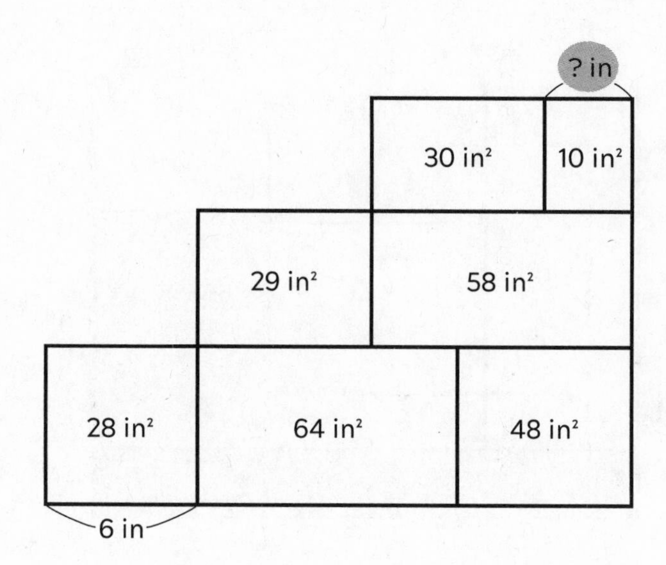

SOLUTION

PUZZLE 83

Find the solution on page 146.

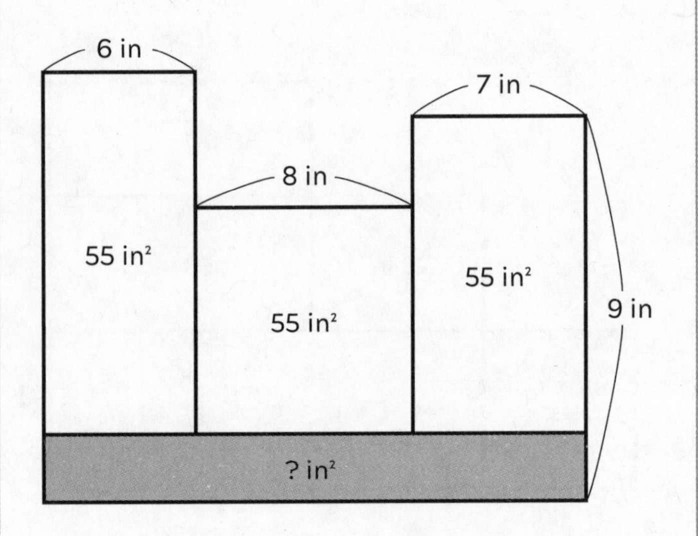

SOLUTION

PUZZLE 84

Find the solution on page 147.

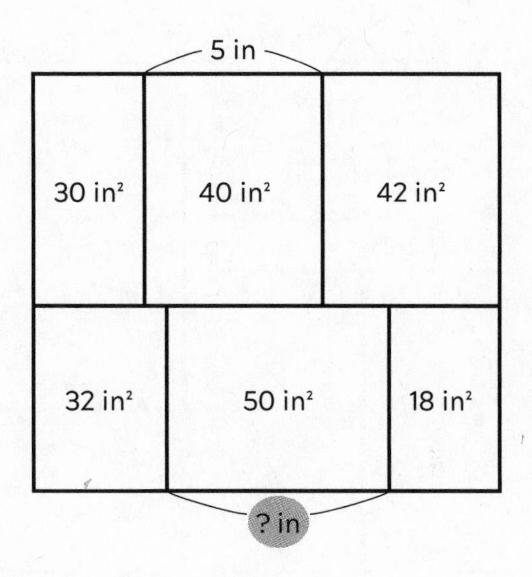

SOLUTION

____

PUZZLE 85

Find the solution on page 147.

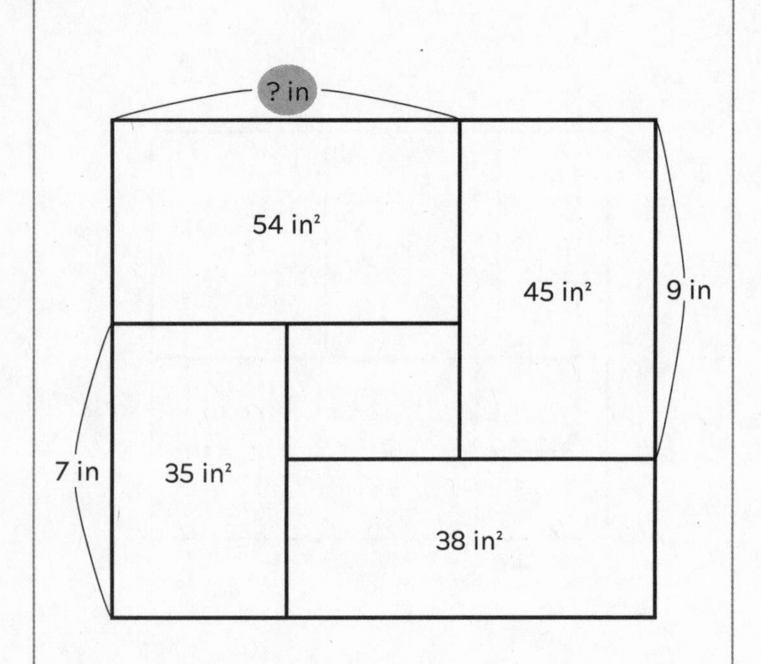

SOLUTION

PUZZLE 86

Find the solution on page 148.

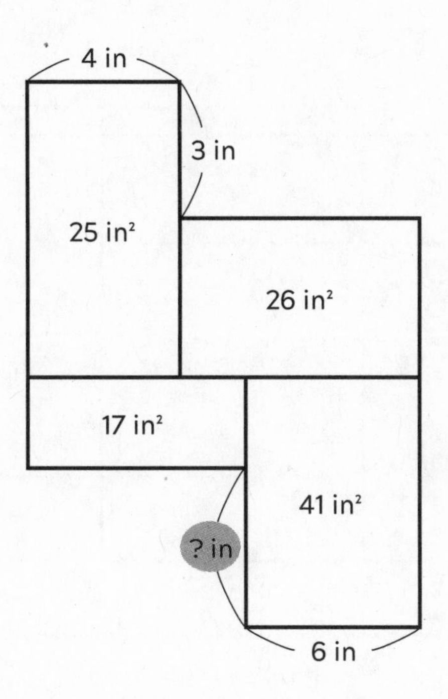

SOLUTION

PUZZLE 87

Find the solution on page 148.

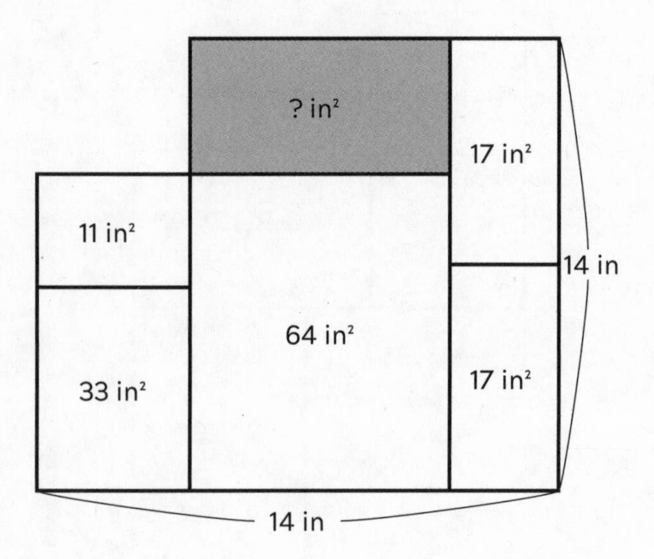

SOLUTION

PUZZLE 88

Find the solution on page 149.

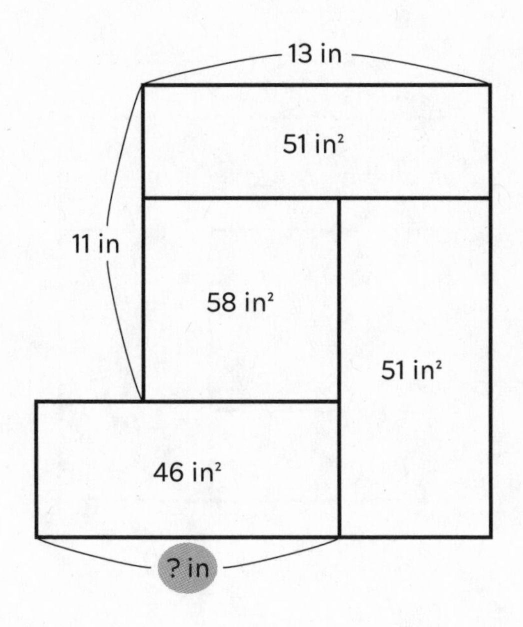

SOLUTION

PUZZLE 89

Find the solution on page 149.

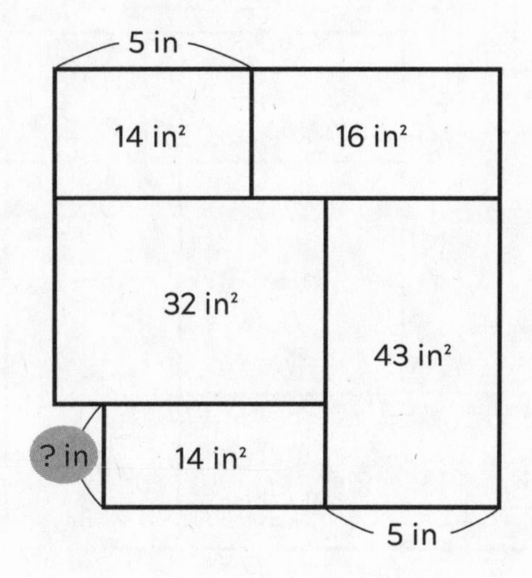

SOLUTION

PUZZLE 90

Find the solution on page 150.

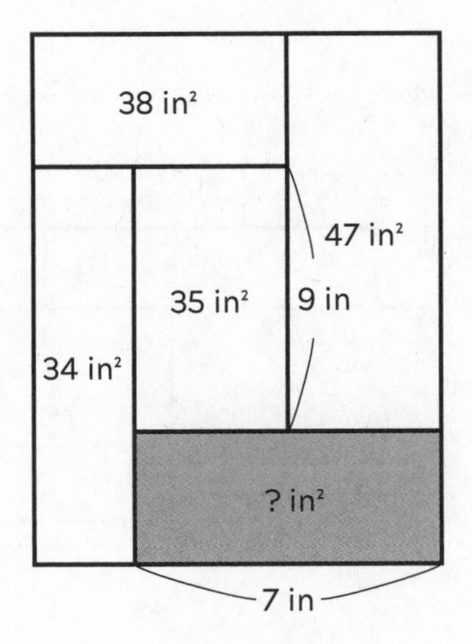

SOLUTION

PUZZLE 91

Find the solution on page 150.

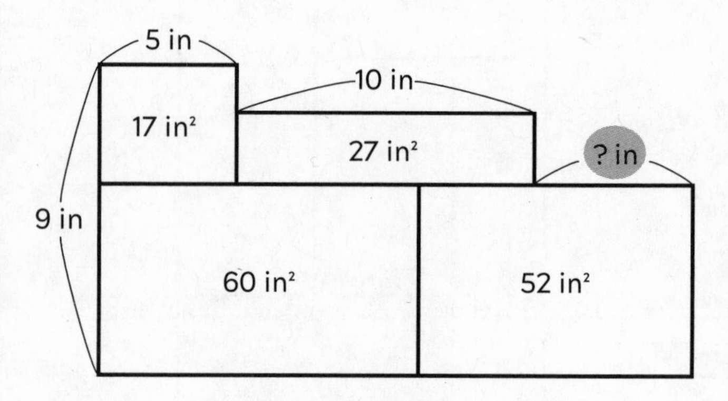

SOLUTION

PUZZLE 92

Find the solution on page 151.

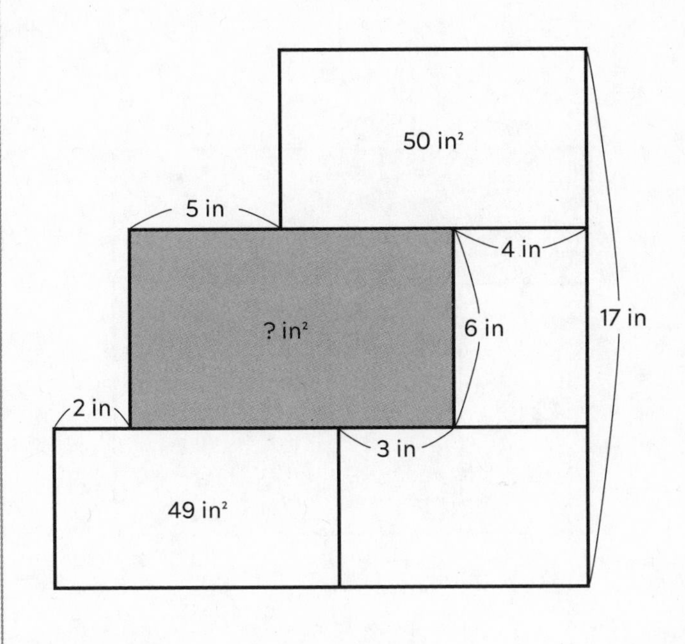

SOLUTION

PUZZLE 93

Find the solution on page 151.

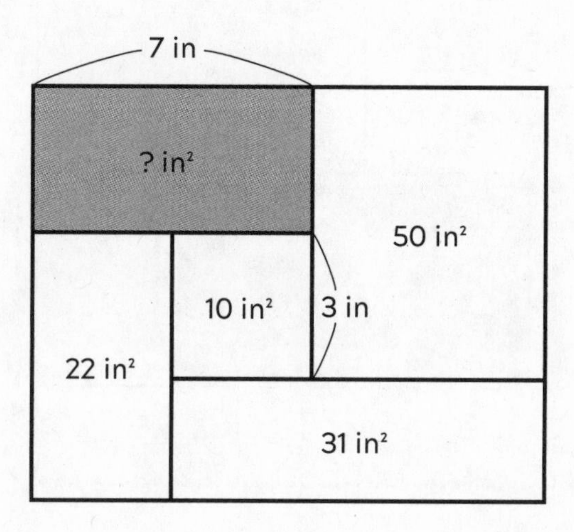

SOLUTION

PUZZLE 94

Find the solution on page 152.

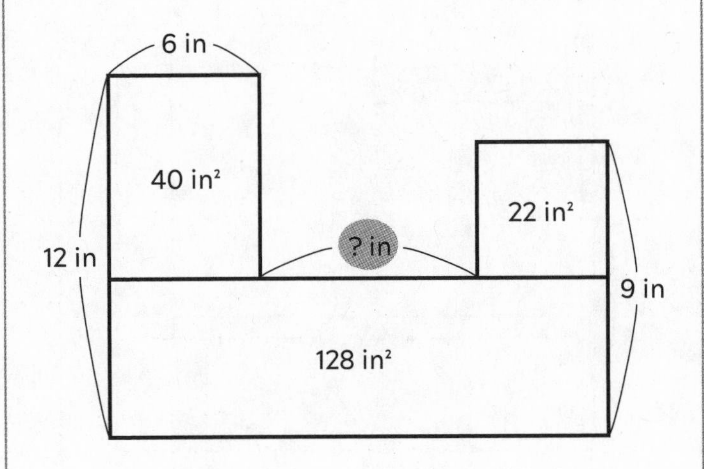

SOLUTION

____

PUZZLE 95

Find the solution on page 152.

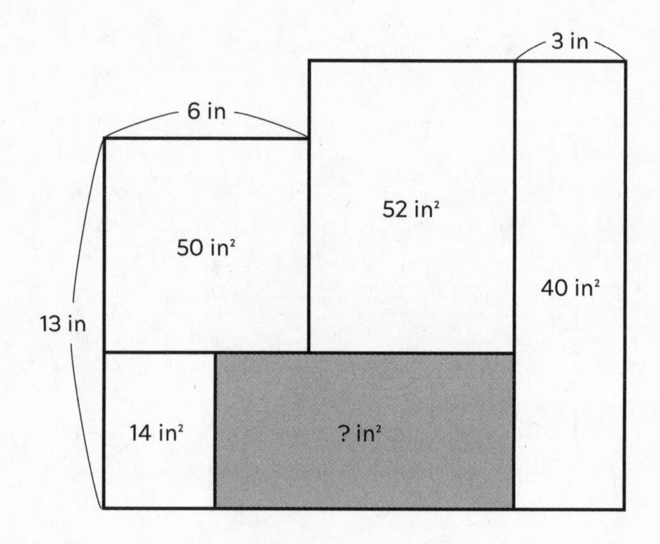

SOLUTION

PUZZLE 96

Find the solution on page 153.

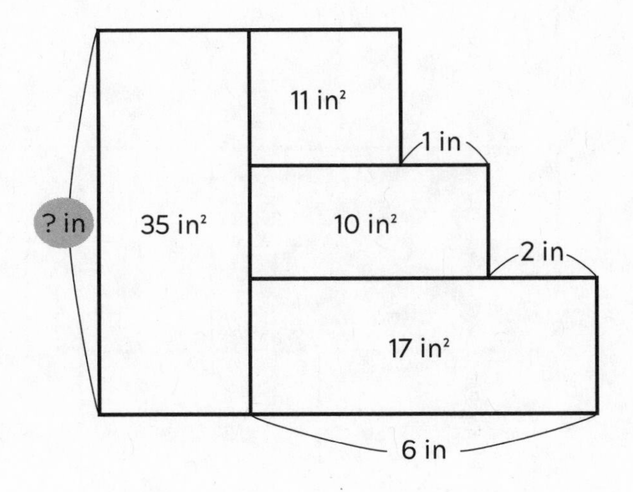

SOLUTION

PUZZLE 97

Find the solution on page 153.

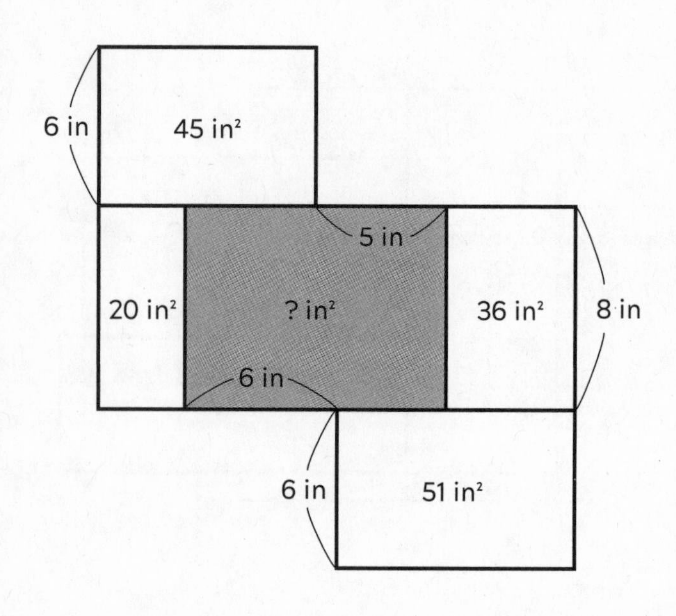

SOLUTION

PUZZLE 98

Find the solution on page 154.

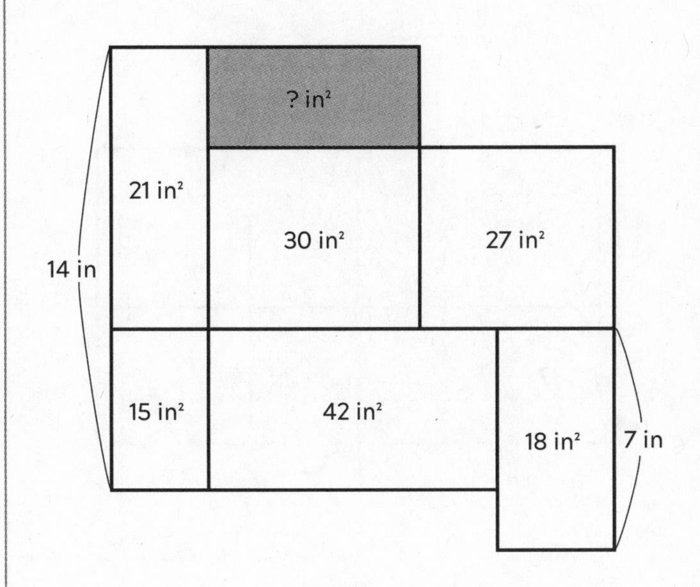

SOLUTION

PUZZLE 99

Find the solution on page 155.

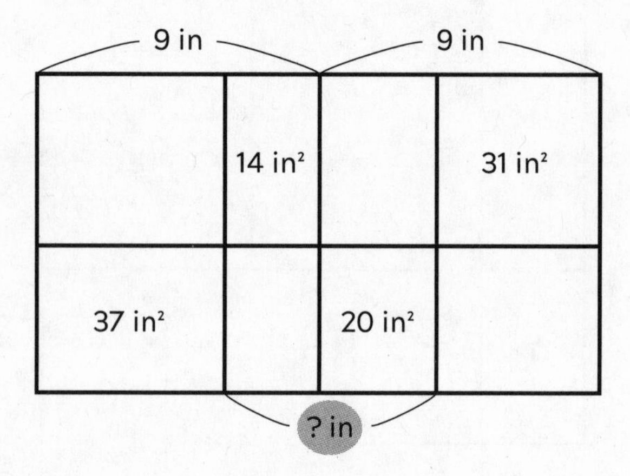

SOLUTION

PUZZLE 100

Find the solution on page 156.

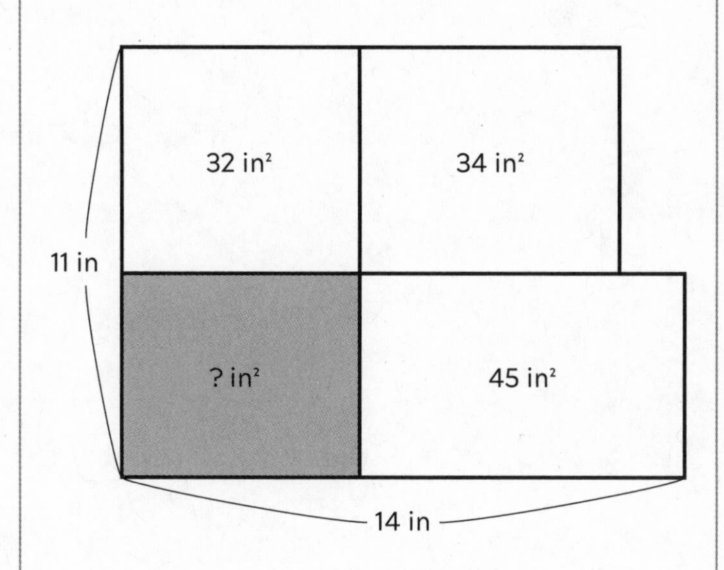

SOLUTION

SOLUTIONS

PUZZLE 1

SOLUTION: 35 in.^{2}

Find length ① . . . $42 \div 6 = 7$ in.

Find length ② . . . $30 \div 6 = 5$ in.

Area ? is $7 \times 5 = 35$ in.^{2}

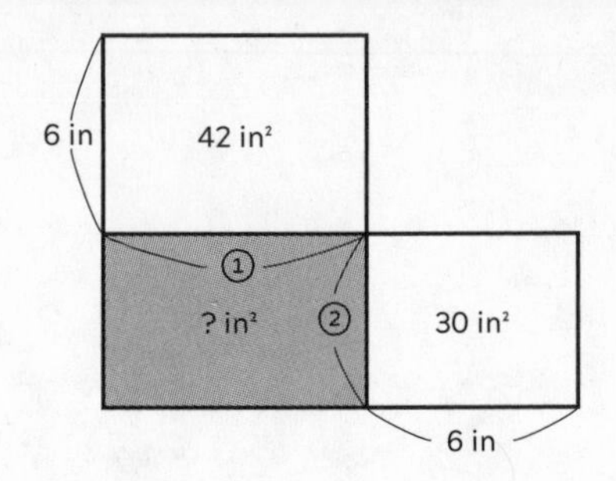

PUZZLE 2

SOLUTION: 16 in.^{2}

Find the total area . . .
$5 \times 8 = 40$ in.^{2}

Area ? is $40 - 24 = 16$ in.^{2}

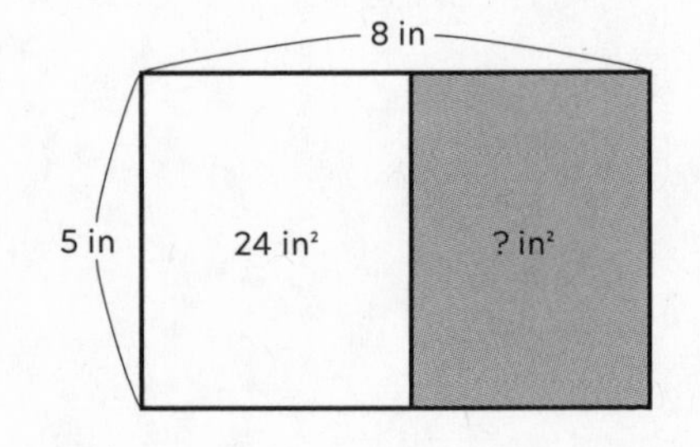

PUZZLE 3

SOLUTION: 6 in.

Find length ① . . . $36 \div 6 = 6$ in.

Find length ② . . . $48 \div 6 = 8$ in.

Find length ③ . . . $64 \div 8 = 8$ in.

Find length ④ . . . $32 \div 8 = 4$ in.

Find length ⑤ . . . $28 \div 4 = 7$ in.

Length ? is $42 \div 7 = 6$ in.

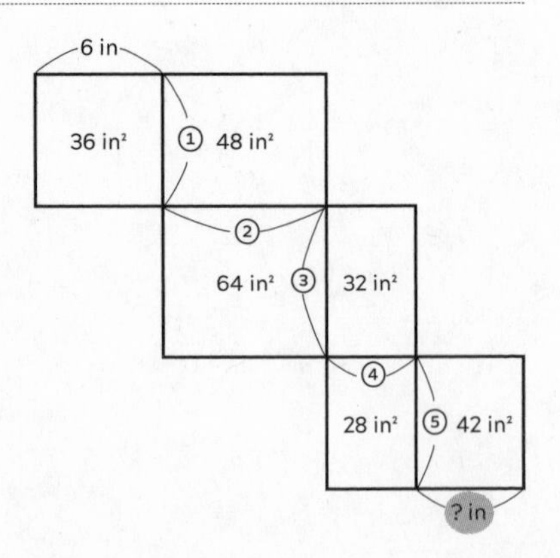

PUZZLE 4

SOLUTION: 24 in.^{2}

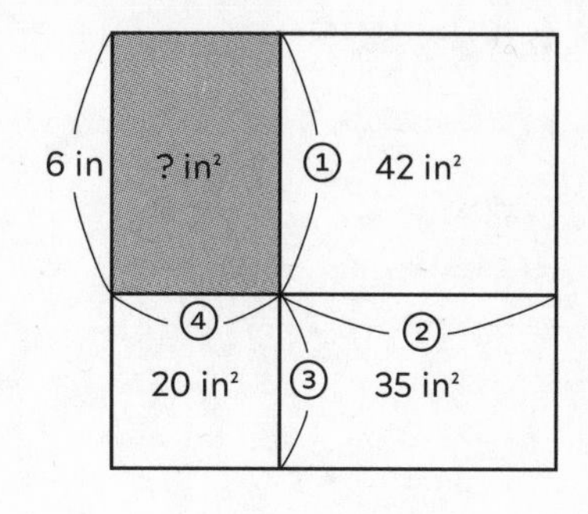

Length ① is 6 in.

Find length ② . . . 42 ÷ 6 = 7 in.

Find length ③ . . . 35 ÷ 7 = 5 in.

Find length ④ . . . 20 ÷ 5 = 4 in.

Area ⓘ is 6 × 4 = 24 in.^{2}

PUZZLE 5

SOLUTION: 12 in.

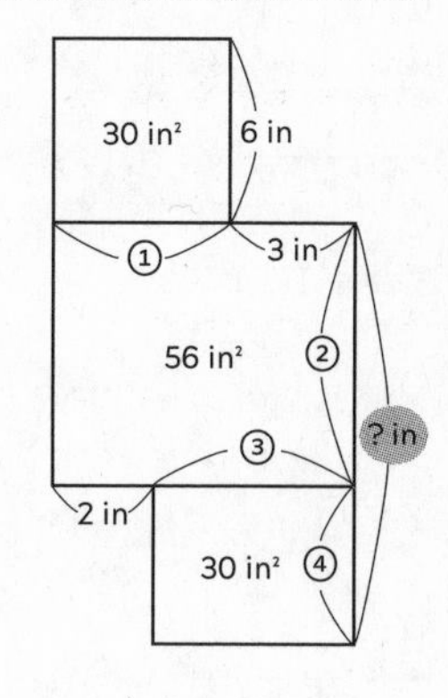

Find length ① . . . 30 ÷ 6 = 5 in.

Find the width of the middle rectangle . . .
5 + 3 = 8 in.

Find length ② . . . 56 ÷ 8 = 7 in.

Find length ③ . . . 8 − 2 = 6 in.

Find length ④ . . . 30 ÷ 6 = 5 in.

Length ? is 7 + 5 = 12 in.

PUZZLE 6

SOLUTION: 9 in.

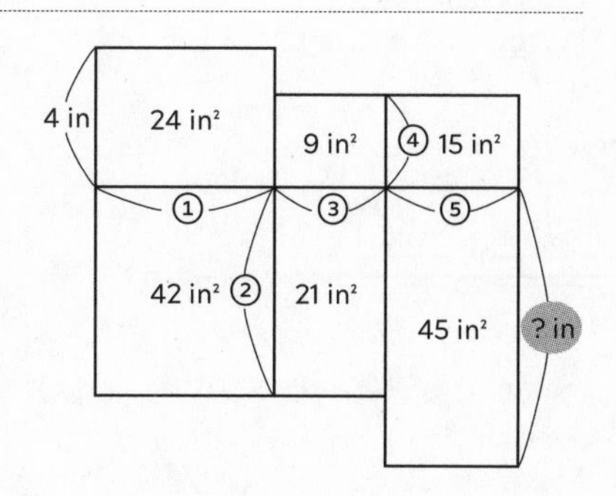

Find length ① . . . 24 ÷ 4 = 6 in.

Find length ② . . . 42 ÷ 6 = 7 in.

Find length ③ . . . 21 ÷ 7 = 3 in.

Find length ④ . . . 9 ÷ 3 = 3 in.

Find length ⑤ . . . 15 ÷ 3 = 5 in.

Length ? is 45 ÷ 5 = 9 in.

PUZZLE 7

SOLUTION: 8 in.

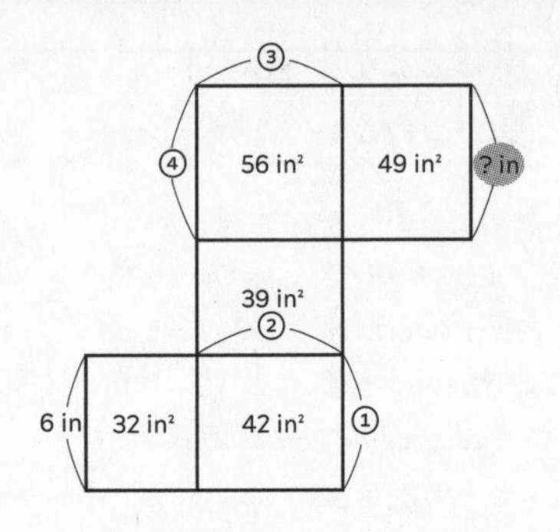

Length ① is 6 in.

Find length ② . . . 42 ÷ 6 = 7 in.

Lengths ② and ③ are the same.

Find length ④ . . . 56 ÷ 7 = 8 in.

Lengths ? and ④ are the same, so length ? is 8 in.

PUZZLE 8

SOLUTION: 30 in.^{2}

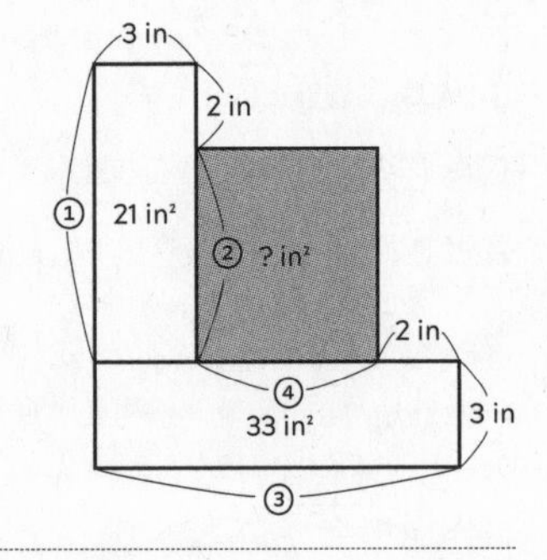

Find length ① . . . 21 ÷ 3 = 7 in.

Find length ② . . . 7 – 2 = 5 in.

Find length ③ . . . 33 ÷ 3 = 11 in.

Find length ④ . . . 11 – 3 – 2 = 6 in.

Area ? is 5 × 6 = 30 in.^{2}

PUZZLE 9

SOLUTION: 5 in.

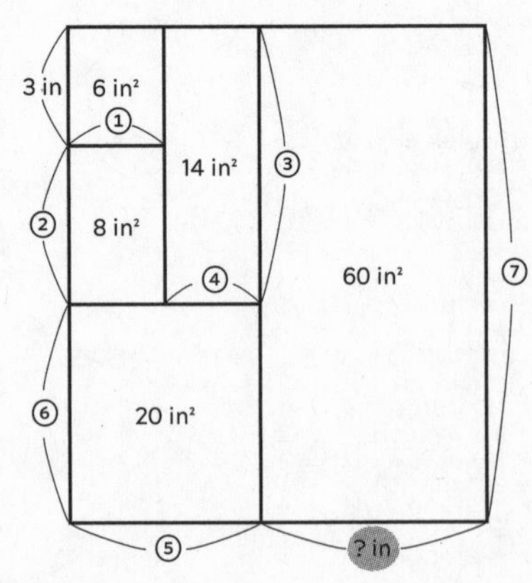

Find length ① . . . 6 ÷ 3 = 2 in.

Find length ② . . . 8 ÷ 2 = 4 in.

Find length ③ . . . 3 + 4 = 7 in.

Find length ④ . . . 14 ÷ 7 = 2 in.

Find length ⑤ . . . 2 + 2 = 4 in.

Find length ⑥ . . . 20 ÷ 4 = 5 in.

Find length ⑦ . . . 7 + 5 = 12 in.

Length ? is 60 ÷ 12 = 5 in.

PUZZLE 10

SOLUTION: 24 in.^{2}

Find length ① . . . 20 ÷ 5 = 4 in.

Find length ② . . . 16 ÷ 4 = 4 in.

Find length ③ . . . 24 ÷ 4 = 6 in.

Find length ④ . . . 35 ÷ 5 = 7 in.

Find length ⑤ . . . 42 ÷ 7 = 6 in.

Find length ⑥ . . . 48 ÷ 6 = 8 in.

Find length ⑦ . . . 32 ÷ 8 = 4 in.

Area ⓘ is 6 × 4 = 24 in.^{2}

PUZZLE 11

SOLUTION: 4 in.

Find length ① . . . 12 ÷ 3 = 4 in.

Find length ② . . . 18 ÷ (3 + 3) = 3 in.

Find length ③ . . . 9 − 4 − 3 = 2 in.

Find length ④ . . . 20 ÷ 2 = 10 in.

Length ⓘ is 10 − (3 + 3) = 4 in.

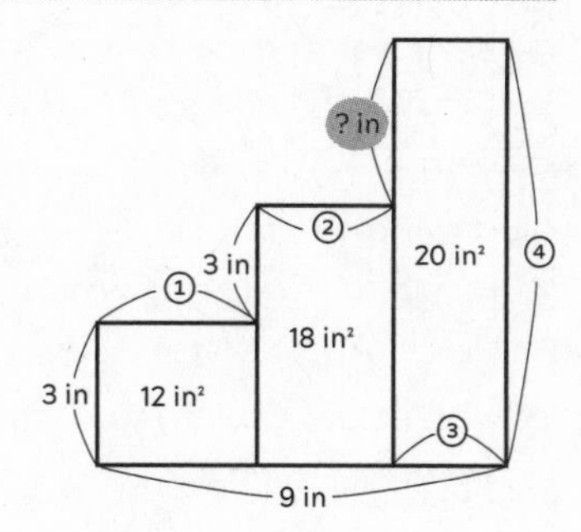

PUZZLE 12

SOLUTION: 36 in.^{2}

Find length ① . . . 48 ÷ 6 = 8 in.

Lengths ① and ② are the same.

Find length ③ . . . 72 ÷ 8 = 9 in.

Find length ④ . . . 36 ÷ 9 = 4 in.

Find length ⑤ . . . 20 ÷ 4 = 5 in.

Find length ⑥ . . . 30 ÷ 5 = 6 in.

Area ⓘ is 6 × 6 = 36 in.^{2}

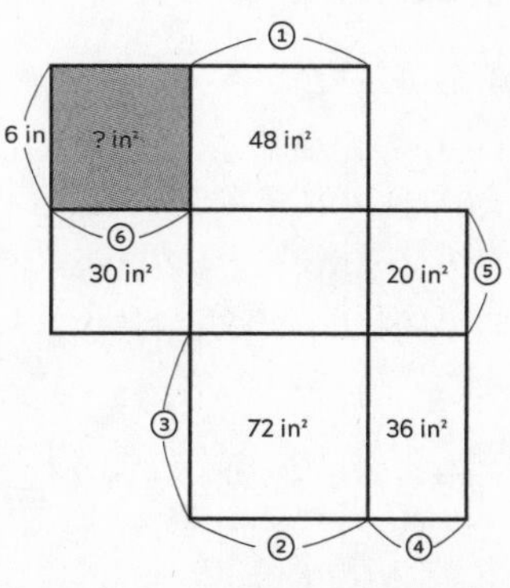

PUZZLE 13

SOLUTION: 35 in.^{2}

Find length ① . . . 48 ÷ 6 = 8 in.

Find length ② . . . 11 – 8 = 3 in.

Find length ③ . . . 27 ÷ 3 = 9 in.

Find length ④ . . . 14 – 9 = 5 in.

Find length ⑤ . . . 14 – 6 = 8 in.

Find length ⑥ . . . 32 ÷ 8 = 4 in.

Find length ⑦ . . . 11 – 4 = 7 in.

Area ? is 5 × 7 = 35 in.^{2}

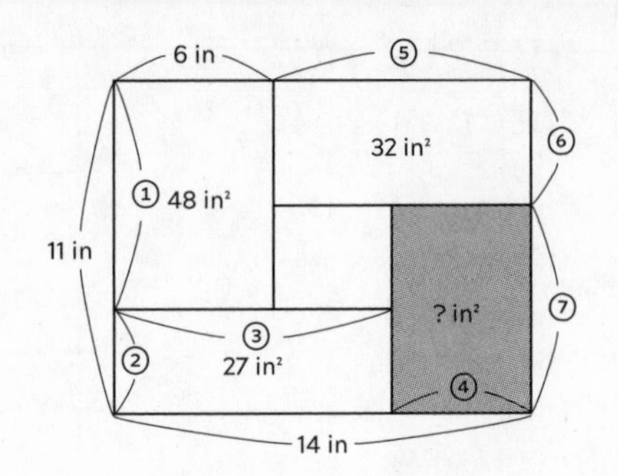

PUZZLE 14

SOLUTION: 5 in.

Find length ① . . . 24 ÷ 4 = 6 in.

Find length ② . . . (6 + 5) – 3 = 8 in.

Find length ③ . . . 8 + 6 = 14 in.

Length ? is 70 ÷ 14 = 5 in.

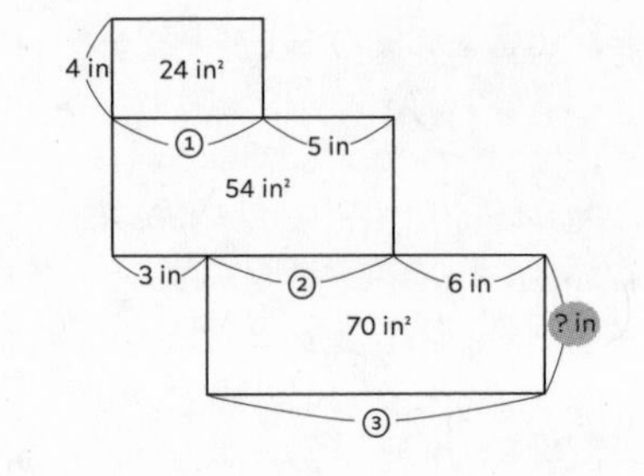

PUZZLE 15

SOLUTION: 12 in.^{2}

Find length ① . . . 9 – 3 = 6 in.

Find length ② . . . 10 – 6 = 4 in.

Find length ③ . . . 7 – 2 = 5 in.

Find length ④ . . . 8 – 5 = 3 in.

Area ? is 4 × 3 = 12 in.^{2}

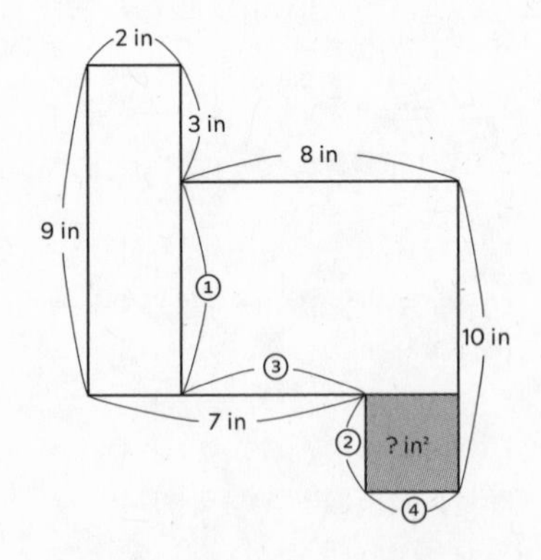

PUZZLE 16

SOLUTION: 2 in.

Find length ① . . . 21 ÷ 7 = 3 in.

Find length ② . . . 4 + 3 = 7 in.

Find length ③ . . . 28 ÷ 7 = 4 in.

Find length ④ . . . 3 + 4 = 7 in.

Find length ⑤ . . . 35 ÷ 7 = 5 in.

Find length ⑥ . . . 5 + 2 = 7 in.

Length ? is 14 ÷ 7 = 2 in.

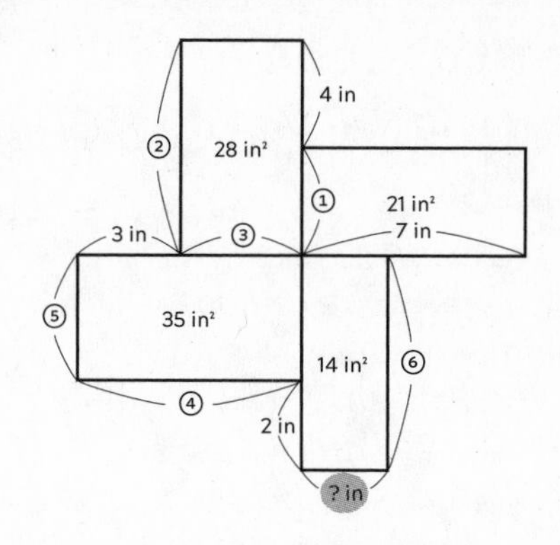

PUZZLE 17

SOLUTION: 4 in.

Find length ① . . . 30 ÷ 6 = 5 in.

Find length ② . . . 35 ÷ 5 = 7 in.

Find length ③ . . . 56 ÷ 7 = 8 in.

Find length ④ . . . 80 ÷ 8 = 10 in.

Find length ⑤ . . . 10 − 6 = 4 in.

Find length ⑥ . . . 36 ÷ 4 = 9 in.

Length ? is 9 − 5 = 4 in.

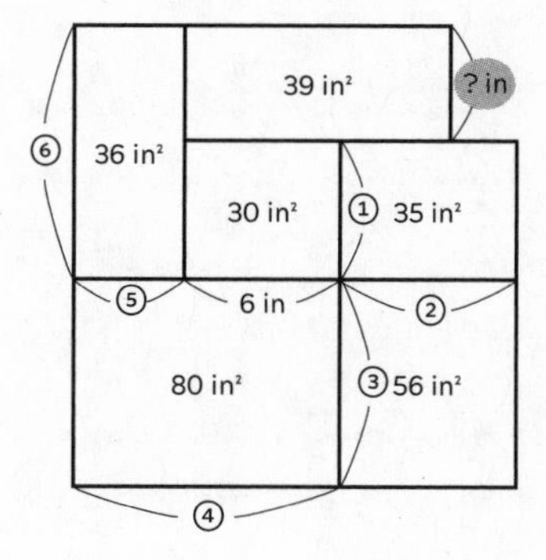

PUZZLE 18

SOLUTION: 30 in.^{2}

Find length ① . . . 40 ÷ 5 = 8 in.

Find length ② . . . 9 ÷ 3 = 3 in.

Find length ③ . . . 8 − 3 = 5 in.

Find length ④ . . . 54 ÷ 6 = 9 in.

Find length ⑤ . . . 12 ÷ 4 = 3 in.

Find length ⑥ . . . 9 − 3 = 6 in.

Area ? is 5 × 6 = 30 in.^{2}

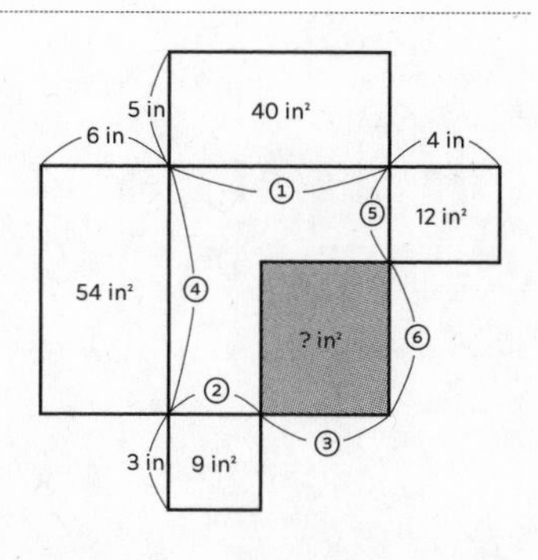

PUZZLE 19

SOLUTION: 17 in.

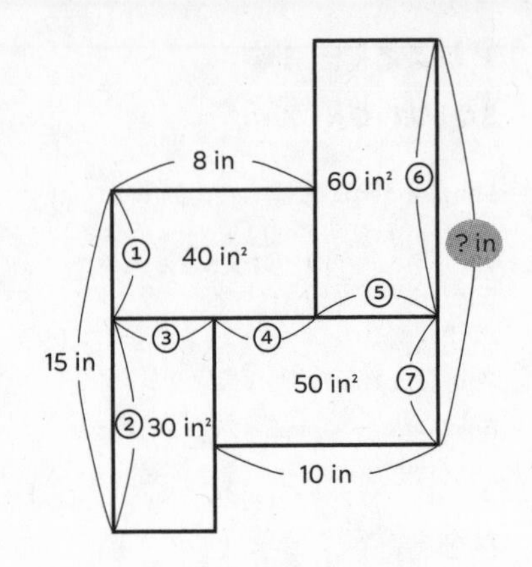

Find length ① . . . 40 ÷ 8 = 5 in.

Find length ② . . . 15 − 5 = 10 in.

Find length ③ . . . 30 ÷ 10 = 3 in.

Find length ④ . . . 8 − 3 = 5 in.

Find length ⑤ . . . 10 − 5 = 5 in.

Find length ⑥ . . . 60 ÷ 5 = 12 in.

Find length ⑦ . . . 50 ÷ 10 = 5 in.

Length ? is 12 + 5 = 17 in.

PUZZLE 20

SOLUTION: 25 in.^{2}

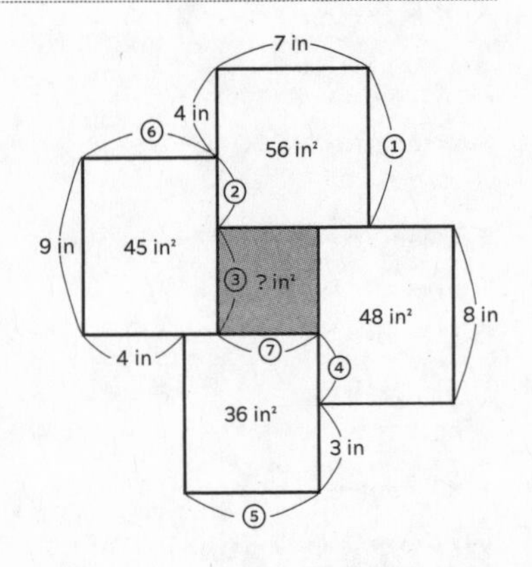

Find length ① . . . 56 ÷ 7 = 8 in.

Find length ② . . . 8 − 4 = 4 in.

Find length ③ . . . 9 − 4 = 5 in.

Find length ④ . . . 8 − 5 = 3 in.

Find length ⑤ . . . 36 ÷ (3 + 3) = 6 in.

Find length ⑥ . . . 45 ÷ 9 = 5 in.

Find length ⑦ . . . 6 − (5 − 4) = 5 in.

Area of ? is 5 × 5 = 25 in.^{2}

PUZZLE 21

SOLUTION: 6 in.

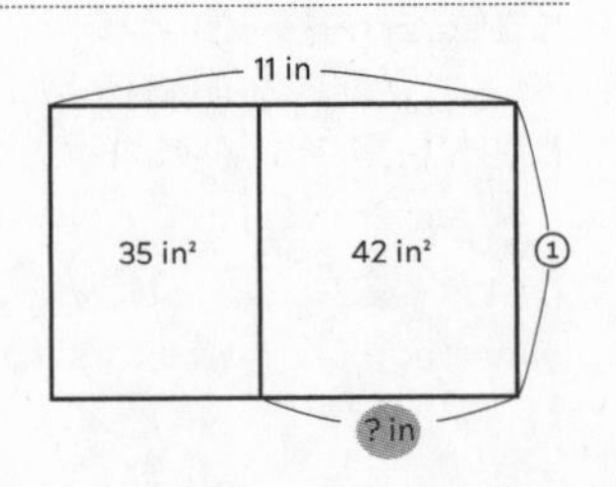

Find the total area . . . 35 + 42 = 77 in.^{2}

Find length ① . . . 77 ÷ 11 = 7 in.

Length ? is 42 ÷ 7 = 6 in.

PUZZLE 22

SOLUTION: 30 in.^{2}

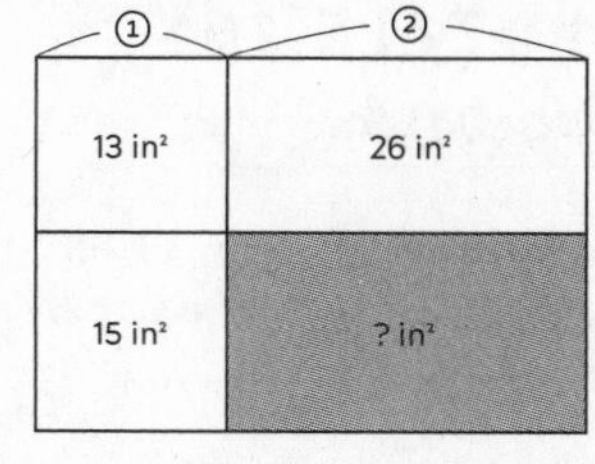

The two rectangles on top are the same height, and 26 in.^{2} is exactly double 13 in.^{2} So, length ② must be exactly double length ①.

Given this, the area on the bottom right must also be double the area on the bottom left.

Area ? is $15 \times 2 = 30$ in.^{2}

PUZZLE 23

SOLUTION: 13 in.^{2}

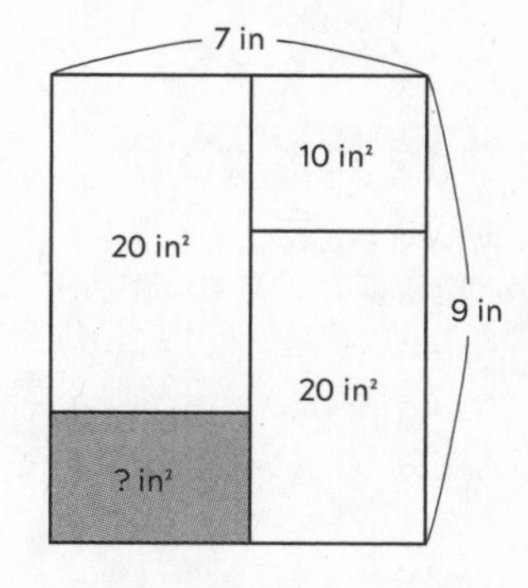

Find the total area . . . $7 \times 9 = 63$ in.^{2}

Subtract the three given areas from the total . . . $63 - (20 + 10 + 20) = 13$ in.^{2}

Area ? is 13 in.^{2}

PUZZLE 24

SOLUTION: 4 in.

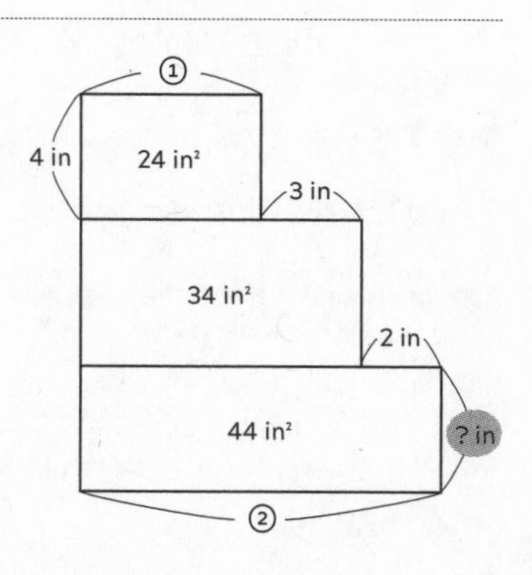

Find length ① . . . $24 \div 4 = 6$ in.

Find length ② . . . $6 + 3 + 2 = 11$ in.

Length ? is $44 \div 11 = 4$ in.

PUZZLE 25

SOLUTION: 5 in.

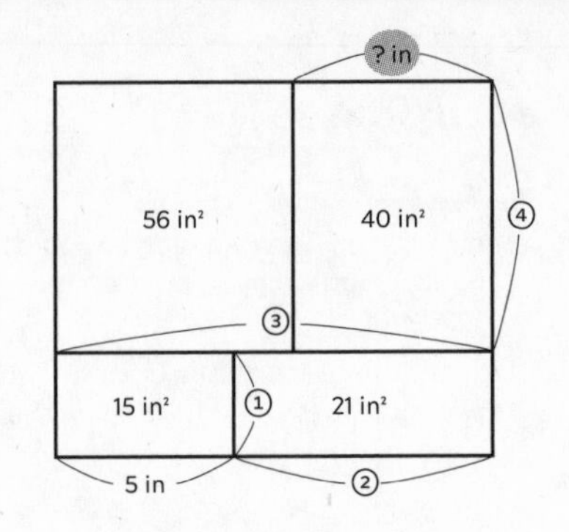

Find length ① . . . 15 ÷ 5 = 3 in.

Find length ② . . . 21 ÷ 3 = 7 in.

Find length ③ . . . 5 + 7 = 12 in.

Add the two areas on top . . .
56 + 40 = 96 in.^{2}

Find length ④ . . . 96 ÷ 12 = 8 in.

Length ? is 40 ÷ 8 = 5 in.

PUZZLE 26

SOLUTION: 81 in.^{2}

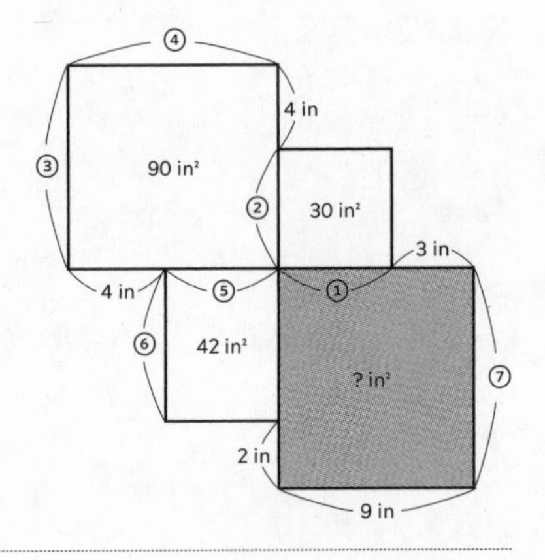

Find length ① . . . 9 − 3 = 6 in.

Find length ② . . . 30 ÷ 6 = 5 in.

Find length ③ . . . 4 + 5 = 9 in.

Find length ④ . . . 90 ÷ 9 = 10 in.

Find length ⑤ . . . 10 − 4 = 6 in.

Find length ⑥ . . . 42 ÷ 6 = 7 in.

Find length ⑦ . . . 7 + 2 = 9 in.

Area ? is 9 × 9 = 81 in.^{2}

PUZZLE 27

SOLUTION: 10 in.

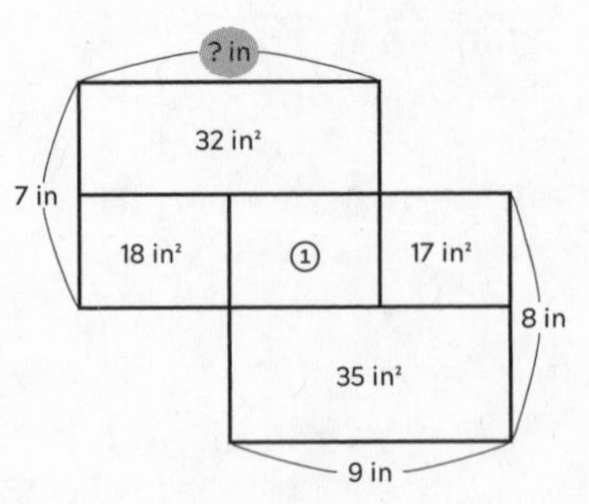

The total area of the three bottom right rectangles is 8 × 9 = 72 in.^{2}

Find area ① . . . 72 − (17 + 35) = 20 in.^{2}

The total area of the three top left rectangles is 32 + 18 + 20 = 70 in.^{2}

Length ? is 70 ÷ 7 = 10 in.

PUZZLE 28

SOLUTION: 5 in.

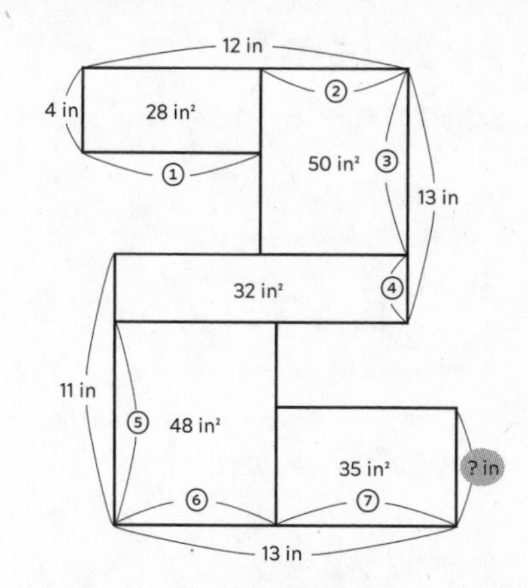

Find length ① . . . 28 ÷ 4 = 7 in.

Find length ② . . . 12 – 7 = 5 in.

Find length ③ . . . 50 ÷ 5 = 10 in.

Find length ④ . . . 13 – 10 = 3 in.

Find length ⑤ . . . 11 – 3 = 8 in.

Find length ⑥ . . . 48 ÷ 8 = 6 in.

Find length ⑦ . . . 13 – 6 = 7 in.

Length ? is 35 ÷ 7 = 5 in.

PUZZLE 29

SOLUTION: 56 in.^{2}

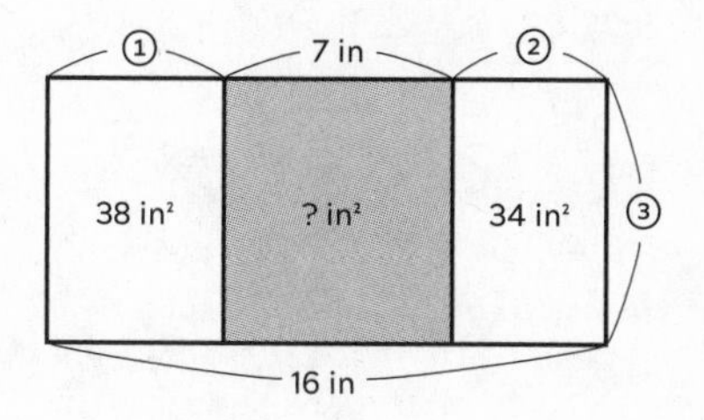

Add the two given areas . . . 38 + 34 = 72 in.^{2}

Their combined width is ① + ②, which is 16 – 7 = 9 in.

Find length ③ . . . 72 ÷ 9 = 8 in.

Area ? is 7 × 8 = 56 in.^{2}

PUZZLE 30

SOLUTION: 4 in.

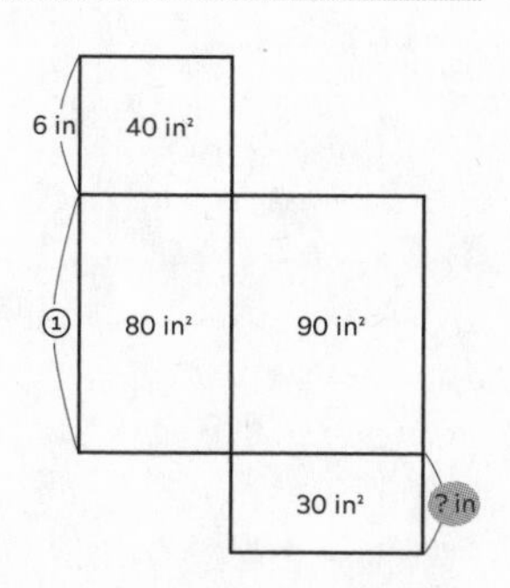

The rectangles on the left are the same width, and 80 in.^{2} is double 40 in.^{2} So, length ① is double 6 in. . . . 6 × 2 = 12 in.

The rectangles on the right are the same width, and 90 in.^{2} is triple 30 in.^{2} So, length ① is triple length ?.

Length ? is 12 ÷ 3 = 4 in.

PUZZLE 31

SOLUTION: 13 in.^{2}

Draw a line to create area ① of 27 in.^{2} Length ② is 8 in.

Find area ③ . . . $42 - 27 = 15$ in.^{2}

Find length ④ . . . $11 - 8 = 3$ in.

Find length ⑤ . . . $15 \div 3 = 5$ in.

Find the total area of the two rectangles on the right . . . $5 \times 8 = 40$ in.^{2}

Area ⓘ is $40 - 27 = 13$ in.^{2}

PUZZLE 32

SOLUTION: 6 in.

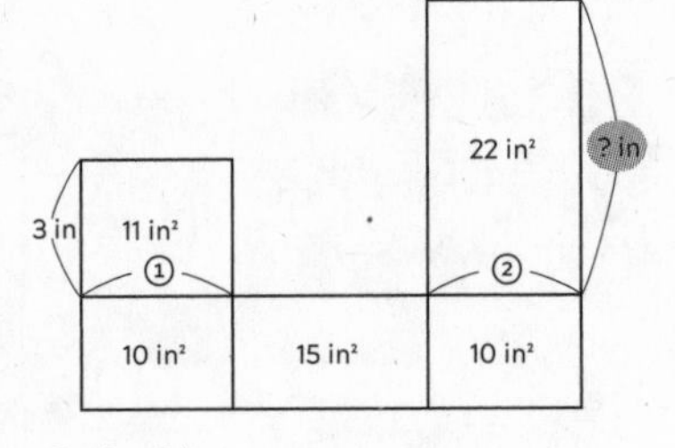

The two 10 in.^{2} rectangles are the same height, so they must be the same width: ① equals ②.

The two rectangles on top are the same width, and 22 in.^{2} is double 11 in.^{2} So, length ⓘ is double 3 in. . . . $3 \times 2 = 6$ in.

PUZZLE 33

SOLUTION: 8 in.

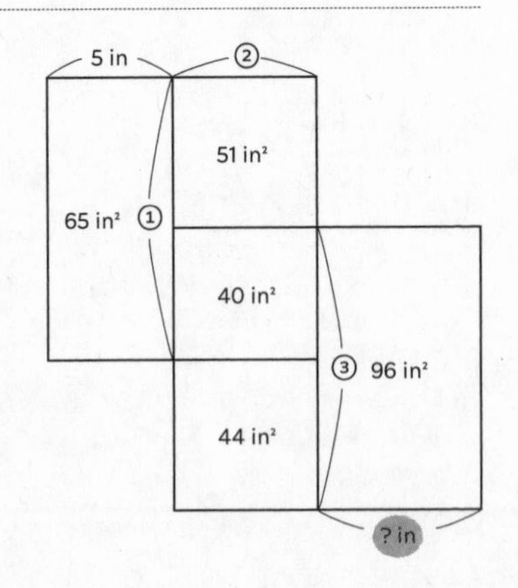

Find length ① . . . $65 \div 5 = 13$ in.

Add the areas with height ① . . . $51 + 40 = 91$ in.^{2}

Find length ② . . . $91 \div 13 = 7$ in.

Add the areas with height ③ . . . $40 + 44 = 84$ in.^{2}

Find length ③ . . . $84 \div 7 = 12$ in.

Length ⓘ is $96 \div 12 = 8$ in.

PUZZLE 34

SOLUTION: 34 in.^{2}

Add the bottom two areas . . .
$59 + 60 = 119$ in.^{2}

Find length ① . . . $119 \div 17 = 7$ in.

Find length ② . . . $12 - 7 = 5$ in.

Find the total area of the top two rectangles . . . $5 \times 14 = 70$ in.^{2}

Area ⓘ is $70 - 36 = 34$ in.^{2}

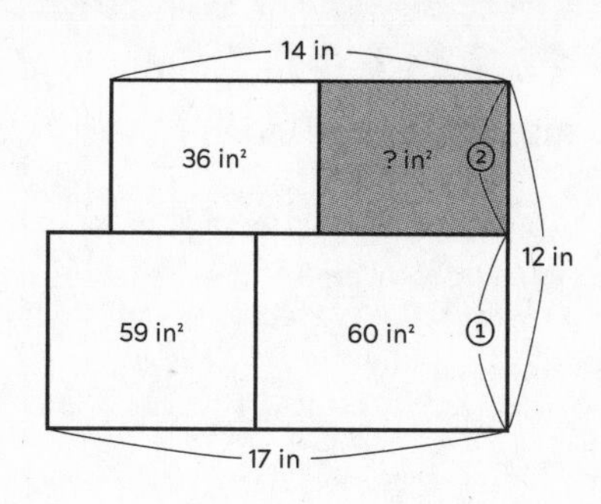

PUZZLE 35

SOLUTION: 12 in.

Draw a line to create areas ① and ②.

Find the total area of the bottom two rectangles . . . $6 \times 10 = 60$ in.^{2}

Find area ① . . . $60 - 32 = 28$ in.^{2}

This is exactly half of 56 in.^{2}, so ① and ② must be identical.

Length ⓘ is $6 \times 2 = 12$ in.

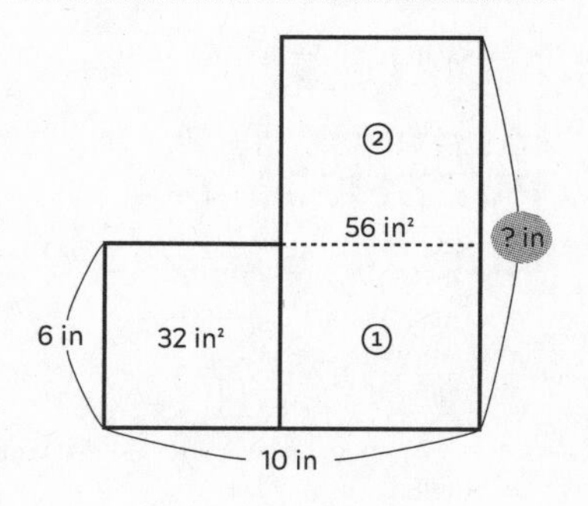

PUZZLE 36

SOLUTION: 21 in.^{2}

Find length ① . . . $24 \div 6 = 4$ in.

Find length ② . . . $28 \div 4 = 7$ in.

Find length ③ . . . $70 \div 7 = 10$ in.

Add the areas with stripes . . .
$28 + 50 = 78$ in.^{2}

Find length ④ . . . $78 \div 6 = 13$ in.

Find length ⑤ . . . $13 - 10 = 3$ in.

Area ⓘ is $7 \times 3 = 21$ in.^{2}

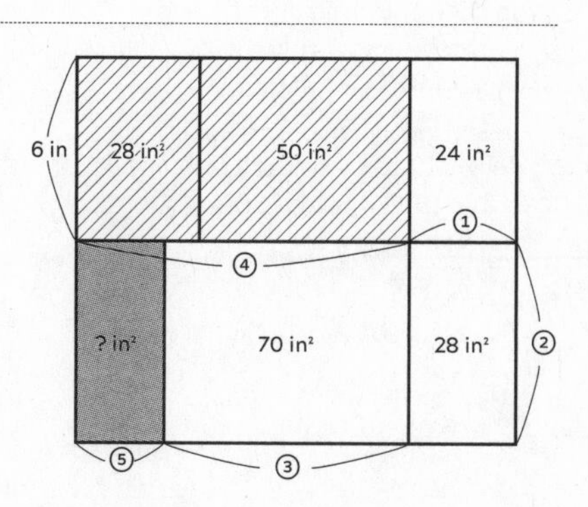

PUZZLE 37

SOLUTION: 30 in.^{2}

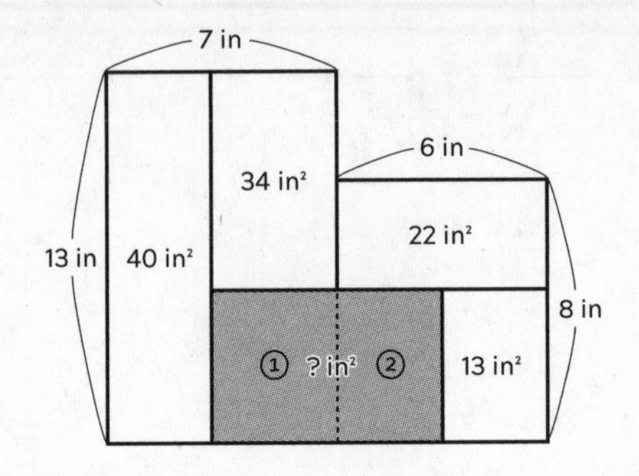

Create areas ① and ②.

Find the total area on the left . . . $13 \times 7 = 91$ in.^{2}

Find area ① . . . $91 - (40 + 34) = 17$ in.^{2}

Find the total area on the right . . . $8 \times 6 = 48$ in.^{2}

Find area ② . . . $48 - (22 + 13) = 13$ in.^{2}

Area ? is $17 + 13 = 30$ in.^{2}

PUZZLE 38

SOLUTION: 15 in.

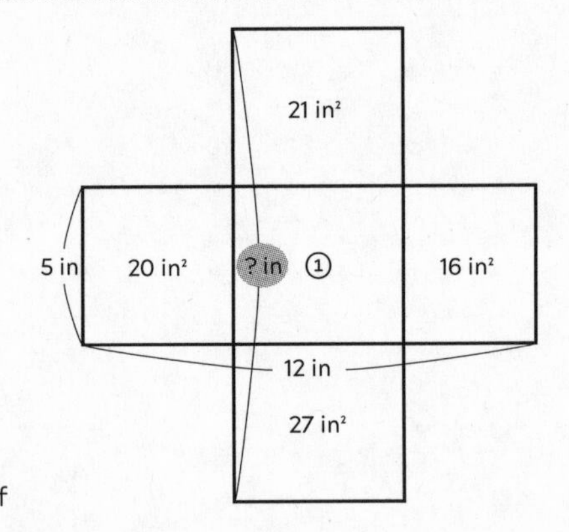

Find the total area of the rectangles in a row . . . $5 \times 12 = 60$ in.^{2}

Find area ① . . . $60 - (20 + 16) = 24$ in.^{2}

Find the total area of the rectangles in a column . . . $21 + 24 + 27 = 72$ in.^{2}

This is exactly triple the area of ①, so length ? must be triple the height of area ①.

Length ? is $5 \times 3 = 15$ in.

PUZZLE 39

SOLUTION: 40 in.^{2}

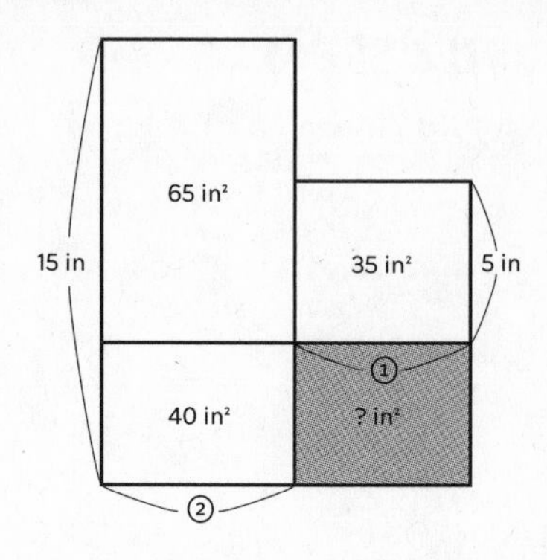

Find length ① . . . 35 ÷ 5 = 7 in.

Add the areas on the left . . .
65 + 40 = 105 in.^{2}

Find length ② . . . 105 ÷ 15 = 7 in.

The rectangles on the bottom have the same height and width, so they are identical.

Area ? is 40 in.^{2}

PUZZLE 40

SOLUTION: 9 in.

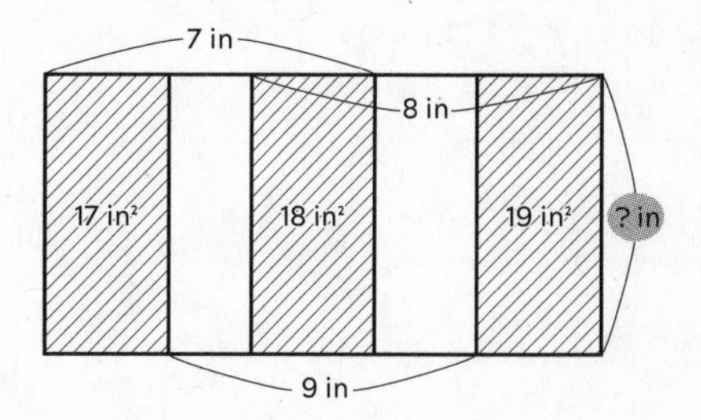

Add the areas with stripes . . .
17 + 18 + 19 = 54 in.^{2}

Find their combined width . . .
7 + 8 – 9 = 6 in.

Length ? is 54 ÷ 6 = 9 in.

PUZZLE 41

SOLUTION: 5 in.

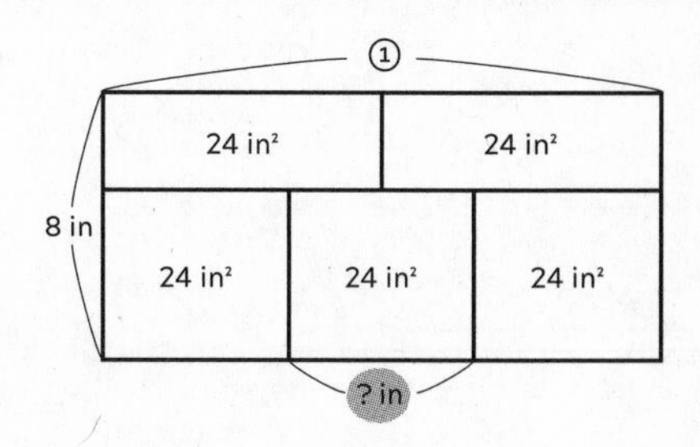

Find the total area . . .
24 × 5 = 120 in.^{2}

Find length ① . . . 120 ÷ 8 = 15 in.

The three rectangles on the bottom have the same area and height, so they must have the same width.

Length ? is 15 ÷ 3 = 5 in.

PUZZLE 42

SOLUTION: 36 in.^{2}

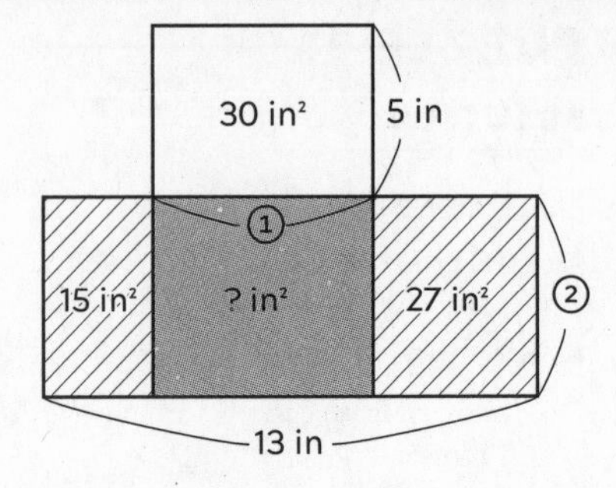

Find length ① . . . 30 ÷ 5 = 6 in.

Add the two areas with stripes . . .
15 + 27 = 42 in.^{2}

Find their combined width . . .
13 – 6 = 7 in.

Find length ② . . . 42 ÷ 7 = 6 in.

Area ⓘ is 6 × 6 = 36 in.^{2}

PUZZLE 43

SOLUTION: 9 in.

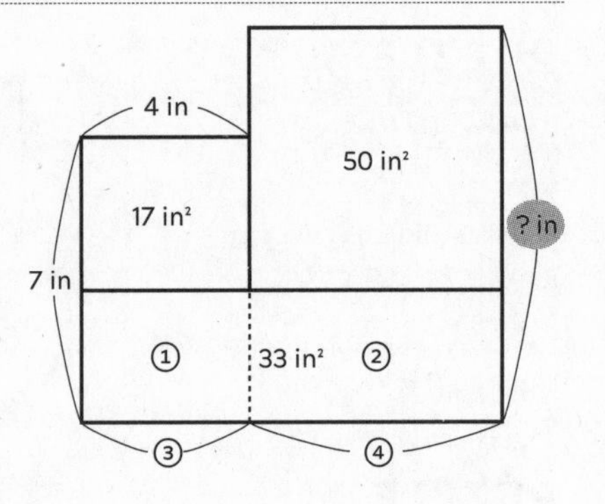

Create areas ① and ②.

Find the total area on the left . . . 7 × 4 = 28 in.^{2}

Find area ① . . . 28 – 17 = 11 in.^{2}

Find area ② . . . 33 – 11 = 22 in.^{2}

Areas ① and ② are the same height, and 22 in.^{2} is double 11 in.^{2} So, length ④ is double length ③.

Find length ④ . . . 4 × 2 = 8 in.

Find the total area on the right . . .
22 + 50 = 72 in.^{2}

Length ⓘ is 72 ÷ 8 = 9 in.

PUZZLE 44

SOLUTION: 4 in.

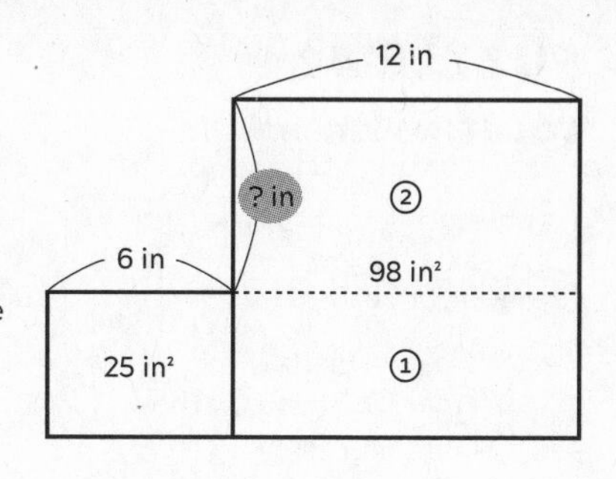

Create areas ① and ②.

Area ① is the same height as the rectangle on the left and twice as wide, so its area must be double 25 in.^{2}

Find area ① . . . 25 × 2 = 50 in.^{2}

Find area ② . . . 98 – 50 = 48 in.^{2}

Length ⓘ is 48 ÷ 12 = 4 in.

PUZZLE 45

SOLUTION: 7 in.

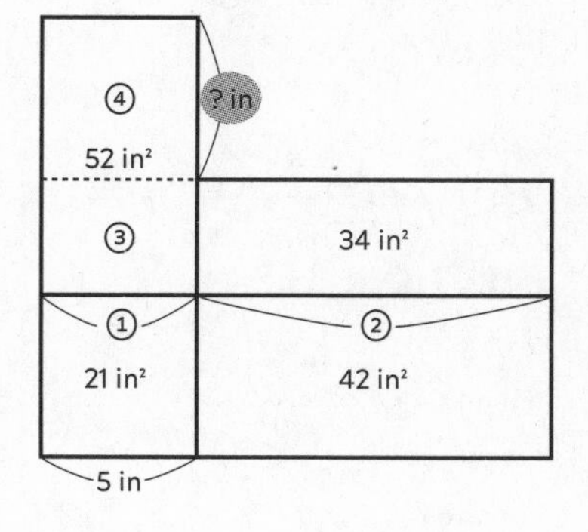

The rectangles on the bottom are the same height, and 42 in.^{2} is double 21 in.^{2} So, length ② must be double length ①.

Create areas ③ and ④.

Area ③ must be half of 34 in.^{2} . . . 34 ÷ 2 = 17 in.^{2}

Find area ④ . . . 52 – 17 = 35 in.^{2}

Length ⓘ is 35 ÷ 5 = 7 in.

PUZZLE 46

SOLUTION: 14 in.

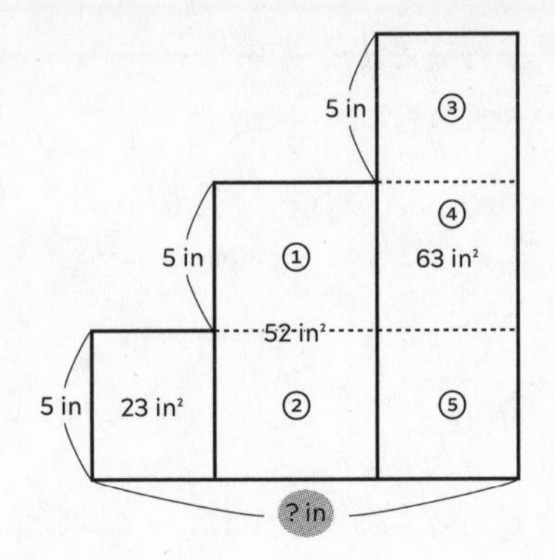

Create areas ① and ②. They are the same height and width, so they must be identical.

Find area ② . . . 52 ÷ 2 = 26 in.^{2}

Create areas ③, ④, and ⑤. They are all the same height and width, so they all must be identical.

Find area ⑤ . . . 63 ÷ 3 = 21 in.^{2}

Add the bottom three areas . . . 23 + 26 + 21 = 70 in.^{2}

Length ⓘ is 70 ÷ 5 = 14 in.

PUZZLE 47

SOLUTION: 8 in.

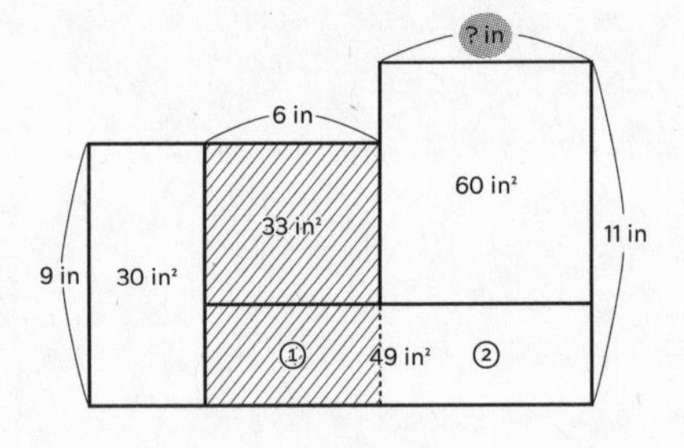

Find the area with stripes . . . 9 × 6 = 54 in.^{2}

Find area ① . . . 54 – 33 = 21 in.^{2}

Find area ② . . . 49 – 21 = 28 in.^{2}

Add the two areas on the right . . . 60 + 28 = 88 in.^{2}

Length ⓘ is 88 ÷ 11 = 8 in.

PUZZLE 48

SOLUTION: 19 in.^{2}

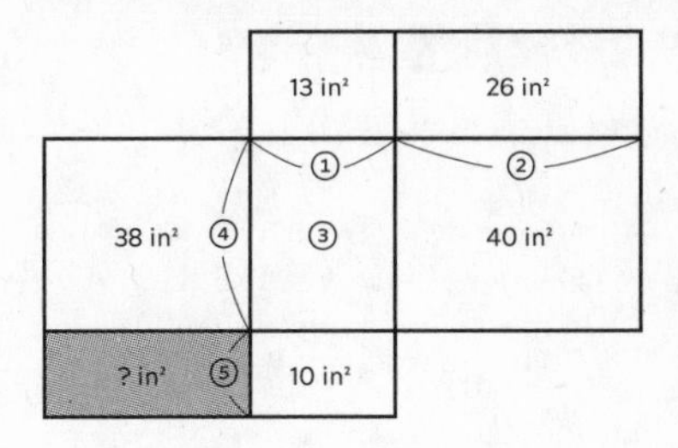

The top two rectangles are the same height, and 26 in.^{2} is double 13 in.^{2} So, length ② must be double length ①.

Area ③ must be half of 40 in.^{2} . . . $40 \div 2 = 20$ in.^{2}

This is double the 10 in.^{2} area below, so length ④ is double length ⑤.

Area ? is half of 38 in.^{2} . . . $38 \div 2 = 19$ in.^{2}

PUZZLE 49

SOLUTION: 4 in.

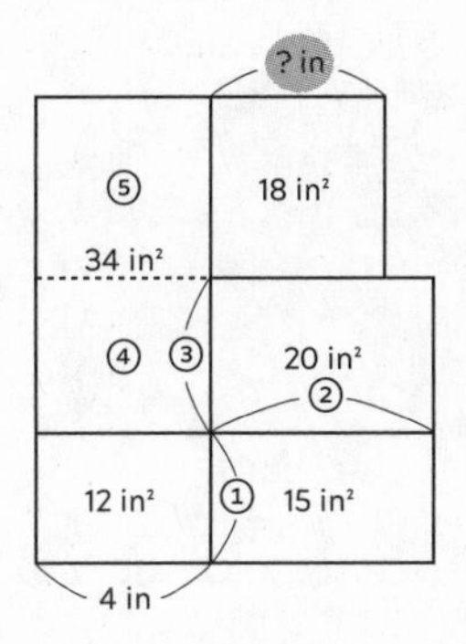

Find length ① . . . $12 \div 4 = 3$ in.

Find length ② . . . $15 \div 3 = 5$ in.

Find length ③ . . . $20 \div 5 = 4$ in.

Create areas ④ and ⑤.

Find area ④ . . . $4 \times 4 = 16$ in.^{2}

Find area ⑤ . . . $34 - 16 = 18$ in.^{2}

Because ⑤ is the same area and height as the rectangle to the right, it must also be the same width.

Length ? is 4 in.

PUZZLE 50

SOLUTION: 16 in.

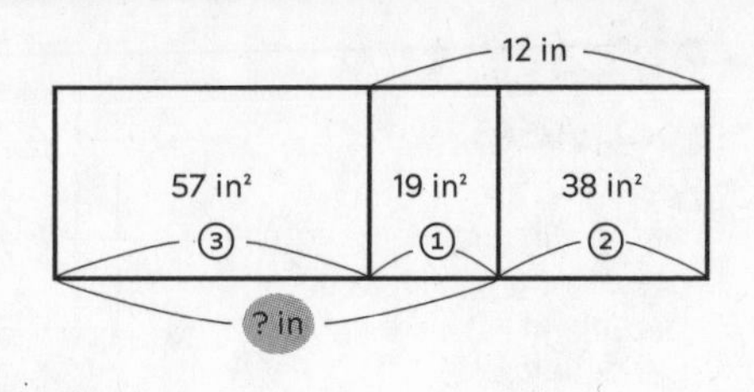

All three rectangles are the same height, and 38 in.² is double 19 in.², while 57 in.² is triple 19 in.²

So, length ② is double length ①, and length ③ is triple length ①.

Find length ① . . . 12 ÷ 3 = 4 in.

Length ? is 4 × 4 = 16 in.

PUZZLE 51

SOLUTION: 15 in.

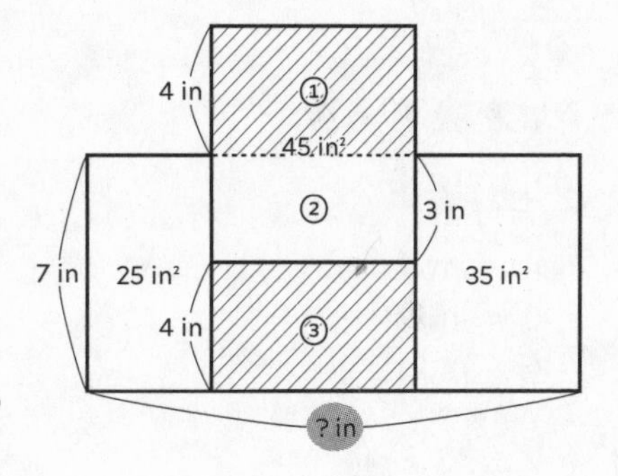

Create areas ① and ②.

Areas ① and ③ have the same height and width, so they must be identical.

Areas ① and ② add to 45 in.², so areas ② and ③ must also add to 45 in.²

Add ②, ③, and the given areas on the right and left . . . 25 + 45 + 35 = 105 in.²

Length ? is 105 ÷ 7 = 15 in.

PUZZLE 52

SOLUTION: 1 in.

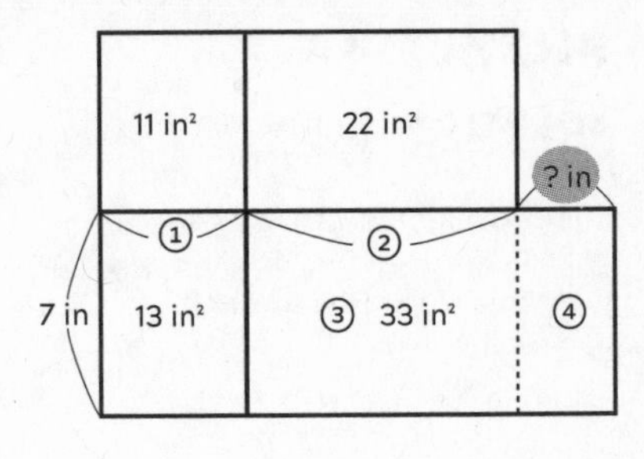

The top two rectangles are the same height, and 22 in.^{2} is double 11 in.^{2} So, length ② is double length ①.

Create areas ③ and ④. Area ③ must be double the area to its left . . . $13 \times 2 = 26$ in.^{2}

Find area ④ . . . $33 - 26 = 7$ in.

Length ⓘ is $7 \div 7 = 1$ in.

PUZZLE 53

SOLUTION: 2 in.

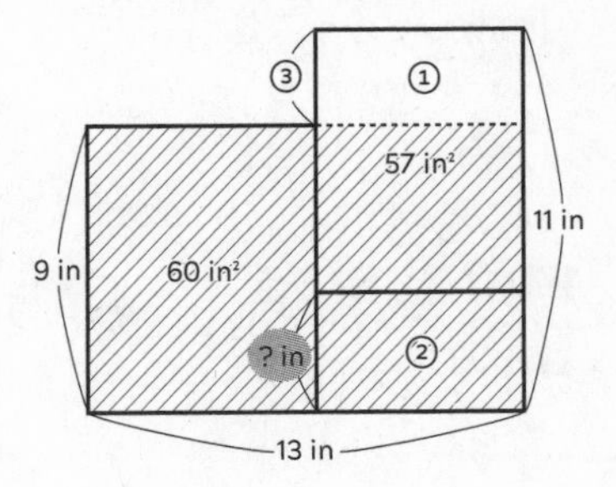

Find the total area with stripes . . . $9 \times 13 = 117$ in.^{2}

This is the same as the sum of the two given areas . . . $60 + 57 = 117$ in.^{2}

So, areas ① and ② must be equal. They are also the same width, so they must be the same height.

Find length ③ . . . $11 - 9 = 2$ in.

Length ? is the same as length ③.

PUZZLE 54

SOLUTION: 12 in.

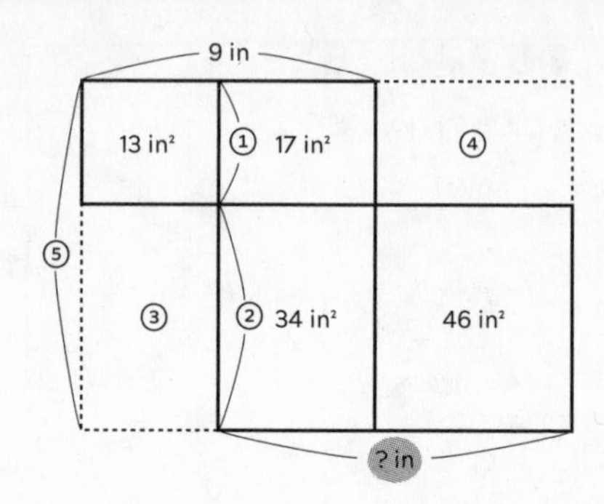

The two middle rectangles are the same width, and 34 in.^{2} is double 17 in.^{2} So, length ② is double length ①.

Create areas ③ and ④.

Find area ③ . . . 13 × 2 = 26 in.^{2}

Find area ④ . . . 46 ÷ 2 = 23 in.^{2}

Add the four areas on the left . . . 13 + 17 + 26 + 34 = 90 in.^{2}

Find length ⑤ . . . 90 ÷ 9 = 10 in.

Add the four areas on the right . . . 17 + 23 + 34 + 46 = 120 in.^{2}

Length ? is 120 ÷ 10 = 12 in.

PUZZLE 55

SOLUTION: 23 in.^{2}

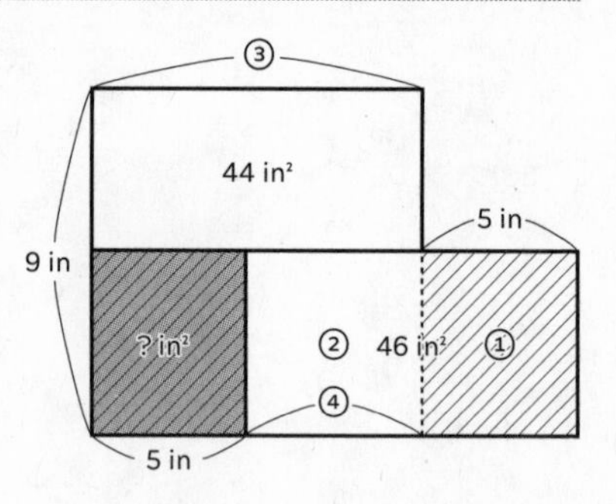

Create areas ① and ②.

Areas ? and ① have the same height and width, so they are identical.

Areas ① and ② add to 46 in.^{2}, so areas ? and ② must also add to 46 in.^{2}

Add the three areas on the left . . . 44 + 46 = 90 in.^{2}

Find length ③ . . . 90 ÷ 9 = 10 in.

Find length ④ . . . 10 − 5 = 5 in.

Area ? must equal area ② . . . 46 ÷ 2 = 23 in.^{2}

PUZZLE 56

SOLUTION: 18 in.^{2}

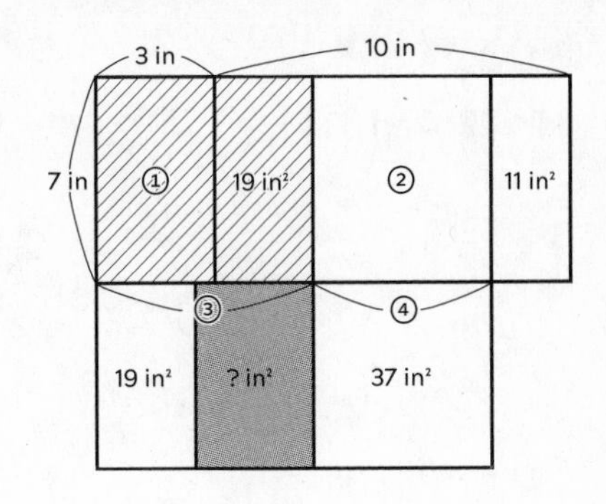

Find area ① . . . $7 \times 3 = 21$ in.^{2}

Find the area with stripes . . .
$21 + 19 = 40$ in.^{2}

Find area ② . . .
$(7 \times 10) - 19 - 11 = 40$ in.^{2}

This is equal to the area with stripes, so length ③ must be equal to length ④.

Area ? plus 19 in.^{2} must equal 37 in.^{2}

Area ? is $37 - 19 = 18$ in.^{2}

PUZZLE 57

SOLUTION: 8 in.^{2}

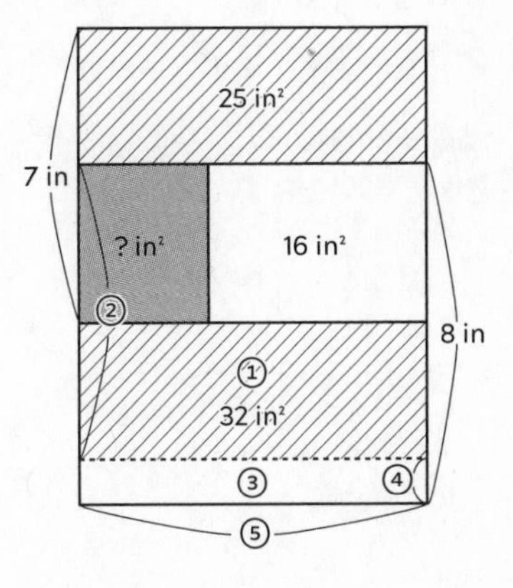

Create area ①, of 25 in.^{2} Length ② is 7 in.

Find area ③ . . . $32 - 25 = 7$ in.^{2}

Find length ④ . . . $8 - 7 = 1$ in.

Find length ⑤ . . . $7 \div 1 = 7$ in.

Find the total area of the top three rectangles . . .
$7 \times 7 = 49$ in.^{2}

Area ? is $49 - (25 + 16) = 8$ in.^{2}

PUZZLE 58

SOLUTION: 8 in.

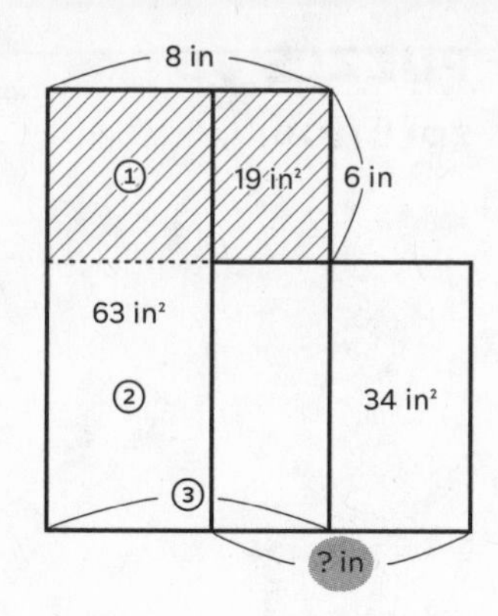

Find the area with stripes . . .
$8 \times 6 = 48$ in.^{2}

Find area ① . . . $48 - 19 = 29$ in.^{2}

Find area ② . . . $63 - 29 = 34$ in.^{2}

The rectangle on the far right and ② have the same height and area, so they must be the same width.

Length ? must be equal to length ③, so length ? is 8 in.

PUZZLE 59

SOLUTION: 42 in.^{2}

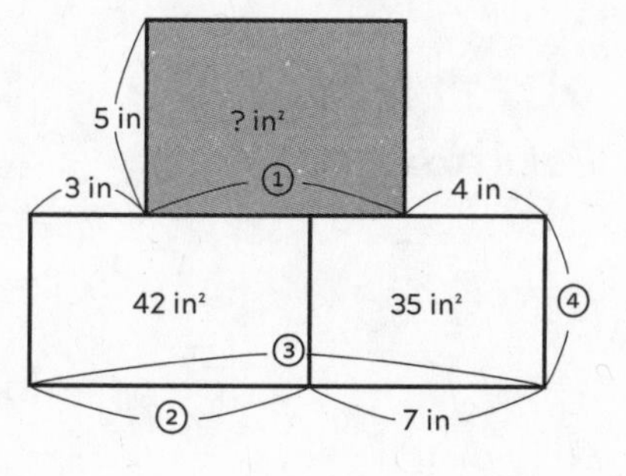

Length ① equals ③ – 3 in. – 4 in., and length ② equals ③ – 7 in.

Because $3 + 4 = 7$, length ① and length ② must be equal.

Find length ④ . . . $35 \div 7 = 5$ in.

The bottom left rectangle and area ? are the same height and width, so they must be identical.

Area ? is 42 in.^{2}

PUZZLE 60

SOLUTION: 29 in.^{2}

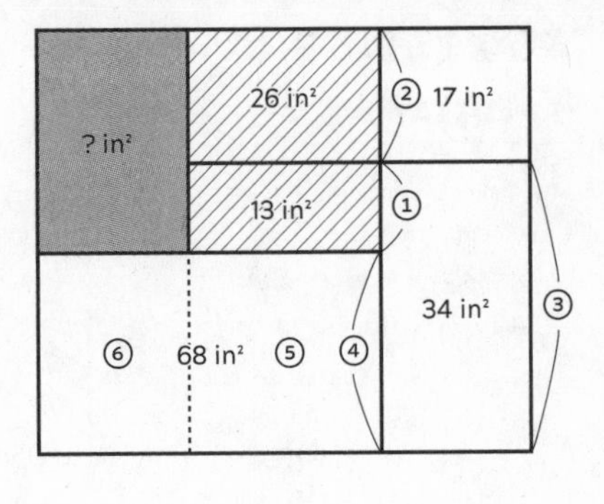

The rectangles with stripes are the same width, and 26 in.^{2} is double 13 in.^{2} So, length ② is double length ①.

The rectangles on the right are the same width, and 34 in.^{2} is double 17 in.^{2} So, length ③ is double length ② (and quadruple length ①).

Length ④ equals ③ – ①, so length ④ is triple length ①.

① + ② also equals triple length ①, so the area with stripes and area ⑤ must be equal . . . 26 + 13 = 39 in.^{2}

Find area ⑥ . . . 68 – 39 = 29 in.^{2}

Area ? is the same as area ⑥.

PUZZLE 61

SOLUTION: 18 in.

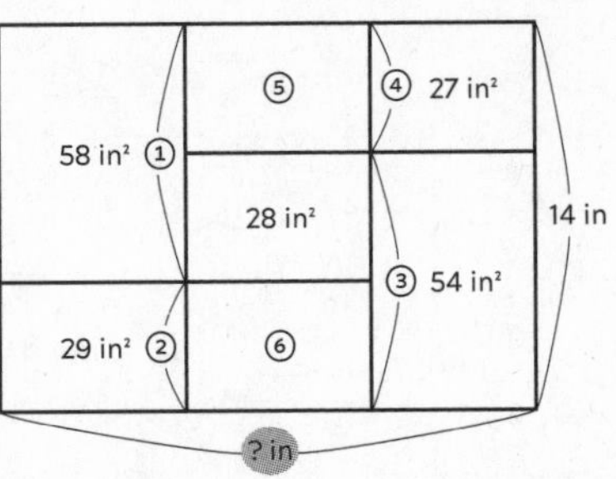

Because 58 in.^{2} is double 29 in.^{2}, length ① is double length ②.

Similarly, length ③ is double length ④.

Based on this, the three rectangles in the middle must all be the same height, so area ⑤ and area ⑥ are both 28 in.^{2}

Find the total area of the diagram . . . 58 + 29 + (28 × 3) + 27 + 54 = 252 in.^{2}

Length ? is 252 ÷ 14 = 18 in.

PUZZLE 62

SOLUTION: 12 in.

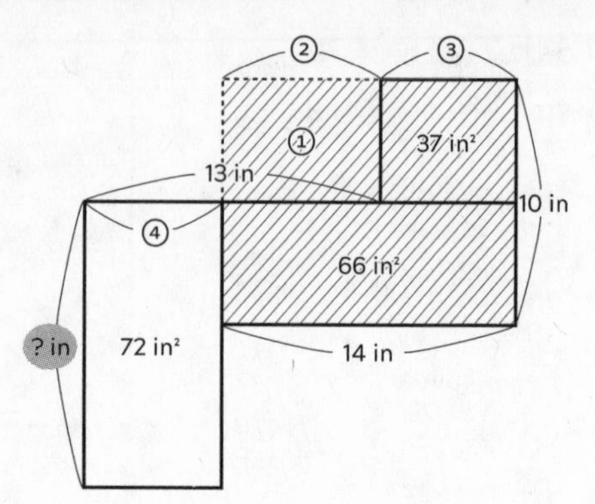

Find the area with stripes . . .
$10 \times 14 = 140$ in.²

Find area ① . . .
$140 - 37 - 66 = 37$ in.²

This is the same area as the top right rectangle, so ② and ③ are the same length.

Find length ② . . . $14 \div 2 = 7$ in.

Find length ④ . . . $13 - 7 = 6$ in.

Length ⑦ is $72 \div 6 = 12$ in.

PUZZLE 63

SOLUTION: 13 in.²

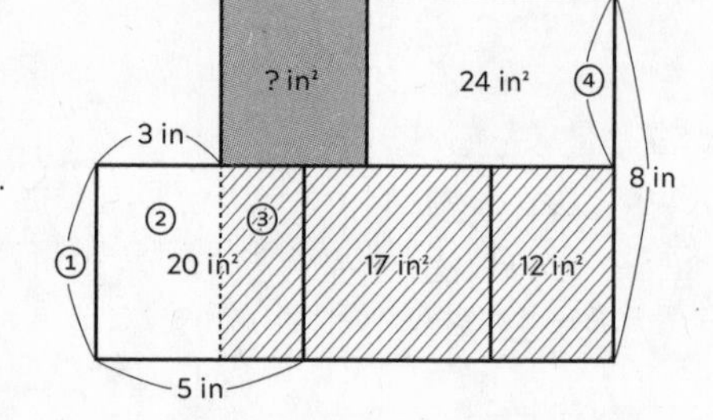

Find length ① . . . $20 \div 5 = 4$ in.

Find area ② . . . $4 \times 3 = 12$ in.²

Find area ③ . . . $20 - 12 = 8$ in.²

Find the area with stripes . . .
$8 + 17 + 12 = 37$ in.²

Find length ④ . . . $8 - 4 = 4$ in.

Because length ① equals length ④, the total area of top two rectangles must equal the area with stripes.

Area ⑦ is $37 - 24 = 13$ in.²

PUZZLE 64

SOLUTION: 12 in.

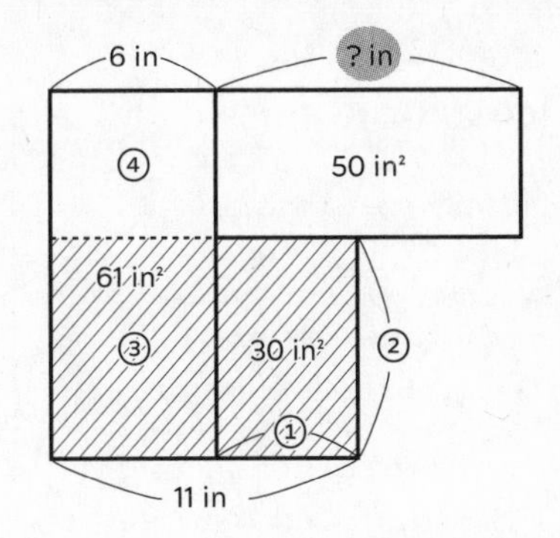

Find length ① . . . 11 – 6 = 5 in.

Find length ② . . . 30 ÷ 5 = 6 in.

Find the area with stripes . . .
$6 \times 11 = 66$ in.^{2}

Find area ③ . . . 66 – 30 = 36 in.^{2}

Find area ④ . . . 61 – 36 = 25 in.^{2}

Because 50 in.^{2} is double 25 in.^{2}, length ? is double 6 in.

Length ? is $6 \times 2 = 12$ in.

PUZZLE 65

SOLUTION: 5 in.

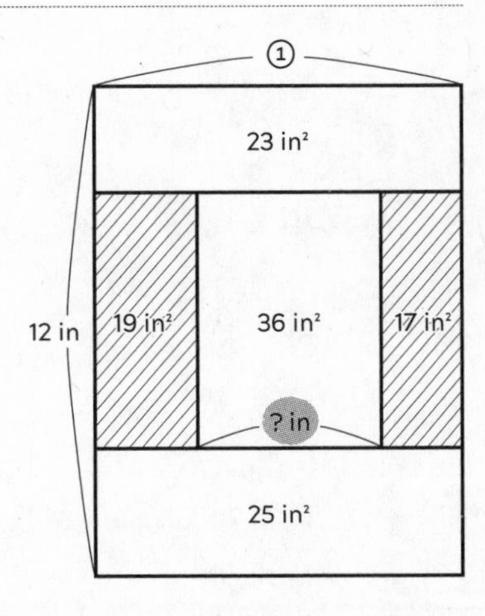

Find the total area . . .
23 + 19 + 36 + 17 + 25 = 120 in.^{2}

Find length ① . . . 120 ÷ 12 = 10 in.

Add the areas with stripes . . .
19 + 17 = 36 in.^{2}

This is the same area as the middle rectangle, so length ? is half of length ①.

Length ? is 10 ÷ 2 = 5 in.

PUZZLE 66

SOLUTION: 4 in.

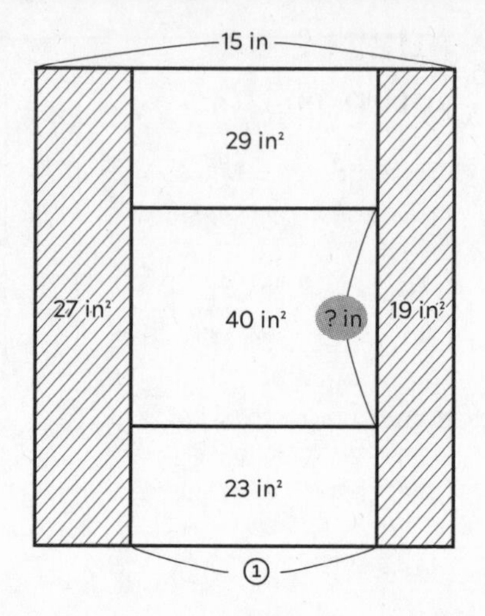

Add the areas with stripes . . .
$27 + 19 = 46$ in.^{2}

Add the areas in the middle . . .
$29 + 40 + 23 = 92$ in.^{2}

Because 92 in.^{2} is double 46 in.^{2}, length ① is double the total width of the areas with stripes.

Find the total width of the areas with stripes . . . $15 \div 3 = 5$ in.

Find length ① . . . $15 - 5 = 10$ in.

Length ? is $40 \div 10 = 4$ in.

PUZZLE 67

SOLUTION: 7 in.

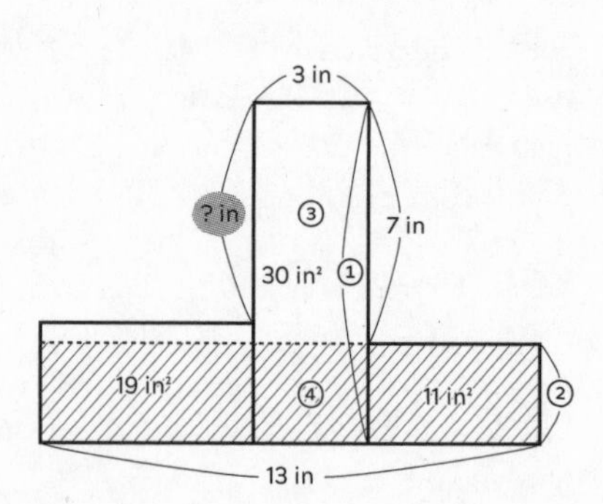

Find length ① . . . $30 \div 3 = 10$ in.

Find length ② . . . $10 - 7 = 3$ in.

Find the area with stripes . . .
$13 \times 3 = 39$ in.^{2}

Create areas ③ and ④.

Find area ③ . . . $3 \times 7 = 21$ in.^{2}

Find area ④ . . . $30 - 21 = 9$ in.^{2}

Subtract area ④ and 11 in.^{2} from the area with stripes . . . $39 - 11 - 9 = 19$ in.^{2}

This is the entire area of the rectangle on the left, so the height of that rectangle must be length ②.

Length ? is 7 in.

PUZZLE 68

SOLUTION: 131 in.^{2}

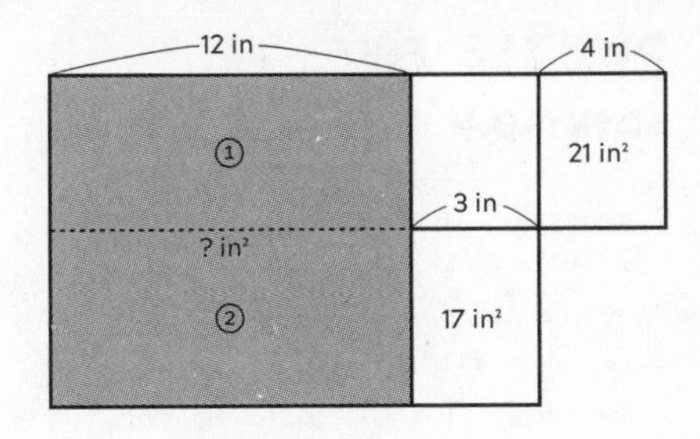

Create areas ① and ②.

Area ① is the same height as the far right rectangle and three times as wide. So it must be three times as big . . . $21 \times 3 = 63$ in.^{2}

Area ② is the same height as the rectangle to its right and four times as wide, so it must be four times as big . . . $17 \times 4 = 68$ in.^{2}

Area ? is $63 + 68 = 131$ in.^{2}

PUZZLE 69

SOLUTION: 24 in.^{2}

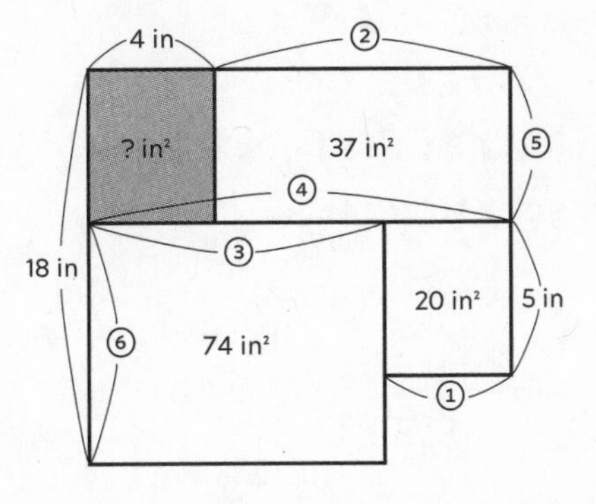

Find length ① . . . $20 \div 5 = 4$ in.

Length ② and length ③ are both equal to ④ – 4 in.

The top right and bottom left rectangles are the same width, and 74 in.^{2} is double 37 in.^{2} So, length ⑥ must be double length ⑤.

Find length ⑤ . . . $18 \div 3 = 6$ in.

Area ? is $4 \times 6 = 24$ in.^{2}

PUZZLE 70

SOLUTION: 3 in.

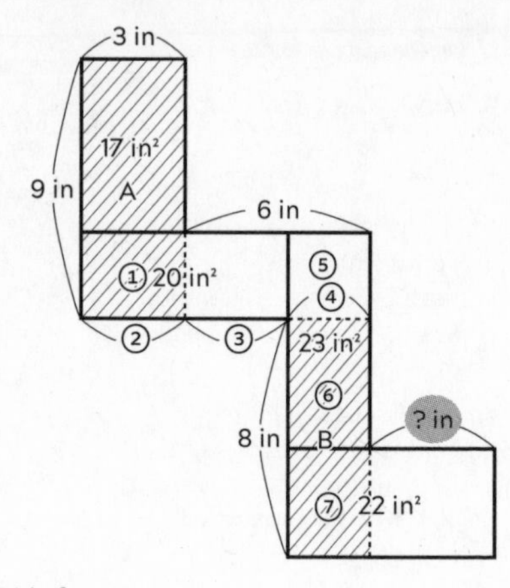

Find area A with stripes . . .
$9 \times 3 = 27$ in.^{2}

Find area ① . . . $27 - 17 = 10$ in.^{2}

This is half of 20 in.^{2}, so length ② and length ③ both equal 3 in.

Find length ④ . . . $6 - 3 = 3$ in.

Area ① and area ⑤ have the same width and height, so area ⑤ must also be 10 in.^{2}

Find area ⑥ . . . $23 - 10 = 13$ in.^{2}

Find area B with stripes . . . $8 \times 3 = 24$ in.^{2}

Find area ⑦ . . . $24 - 13 = 11$ in.^{2}

This is half of 22 in.^{2}, so length ? must equal length ④.

Length ? is 3 in.

PUZZLE 71

SOLUTION: 44 in.^{2}

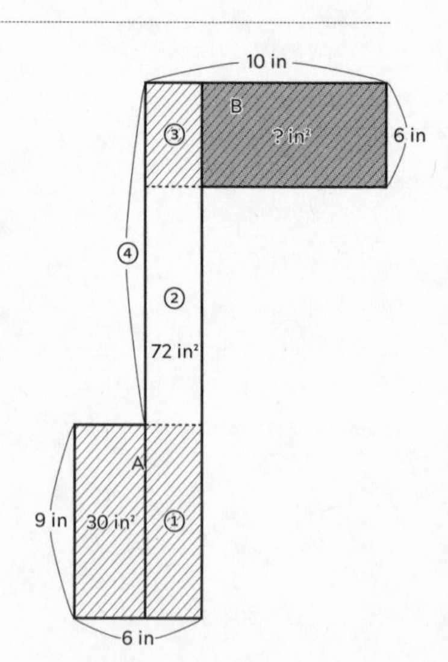

Find area A with stripes . . . $9 \times 6 = 54$ in.^{2}

Find area ① . . . $54 - 30 = 24$ in.^{2}

Find the combined area of ② and ③ . . .
$72 - 24 = 48$ in.^{2}

This is double area ①, so length ④ must be double 9 in. . . . $9 \times 2 = 18$ in.

This is triple 6 in., so area ③ is . . .
$48 \div 3 = 16$ in.^{2}

Find area B with stripes . . .
$6 \times 10 = 60$ in.^{2}

Area ? is $60 - 16 = 44$ in.^{2}

PUZZLE 72

SOLUTION: 29 in.^{2}

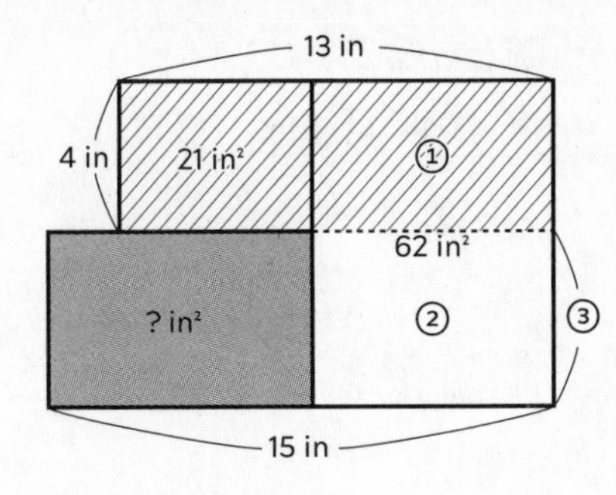

Find the area with stripes . . .
$4 \times 13 = 52$ in.^{2}

Find area ① . . . $52 - 21 = 31$ in.^{2}

Find area ② . . . $62 - 31 = 31$ in.^{2}

Areas ① and ② are equal and have the same width, so they are identical and length ③ equals 4 in.

Find the area without stripes . . . $4 \times 15 = 60$ in.^{2}

Area ? is $60 - 31 = 29$ in.^{2}

PUZZLE 73

SOLUTION: 15 in.

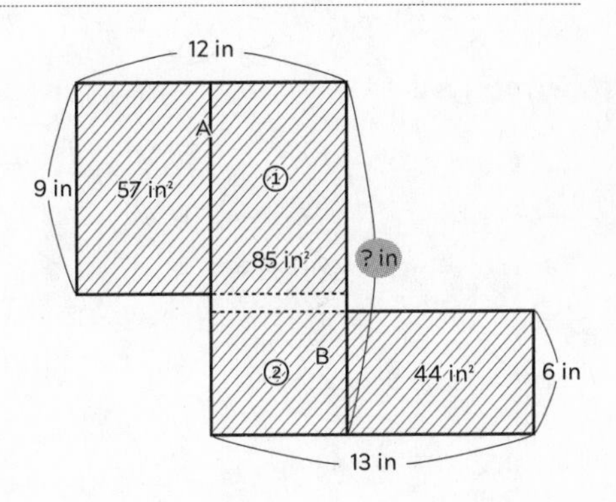

Find area A with stripes . . .
$9 \times 12 = 108$ in.^{2}

Find area ① . . .
$108 - 57 = 51$ in.^{2}

Find area B with stripes . . .
$13 \times 6 = 78$ in.^{2}

Find area ② . . .
$78 - 44 = 34$ in.^{2}

Add areas ① and ② . . .
$51 + 34 = 85$ in.^{2}

This is the entire area of the middle rectangle, so there is no gap between areas ① and ②.

Length ? is $9 + 6 = 15$ in.

PUZZLE 74

SOLUTION: 20 in.

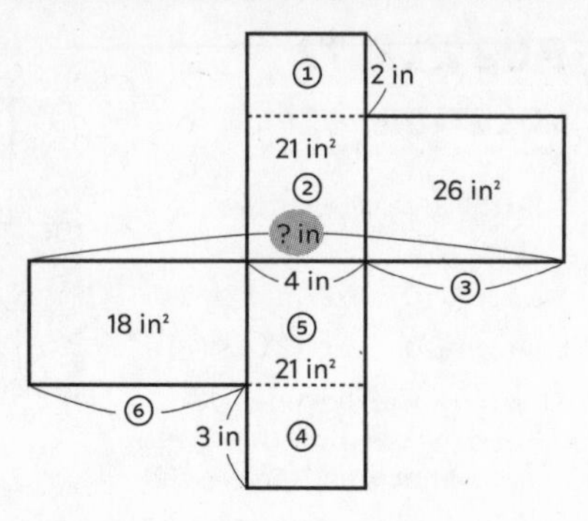

Find area ① . . . 2 × 4 = 8 in.²

Find area ② . . . 21 – 8 = 13 in.²

This is half of 26 in.² So, length ③ is double 4 in. . . . 4 × 2 = 8 in.

Find area ④ . . . 3 × 4 = 12 in.²

Find area ⑤ . . . 21 – 12 = 9 in.²

This is half of 18 in.² So, length ⑥ is double 4 in. . . . 4 × 2 = 8 in.

Length ⑦ is 8 + 4 + 8 = 20 in.

PUZZLE 75

SOLUTION: 4 in.

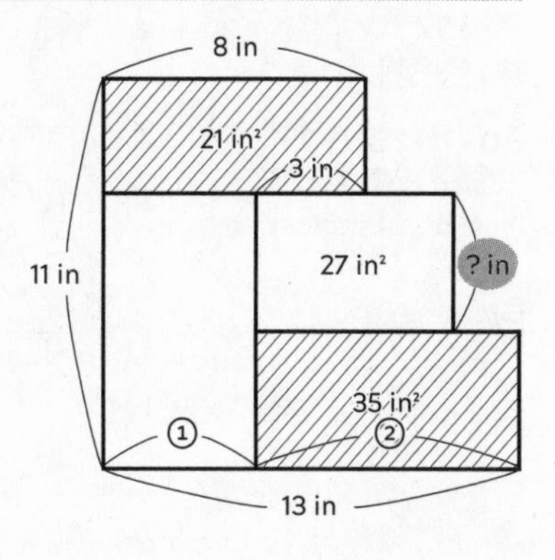

Find length ① . . . 8 – 3 = 5 in.

Find length ② . . . 13 – 5 = 8 in.

The two rectangles with stripes are the same width. Add their areas . . . 21 + 35 = 56 in.²

Find their combined height . . . 56 ÷ 8 = 7 in.

Length ⑦ is 11 – 7 = 4 in.

PUZZLE 76

SOLUTION: 46 in.^{2}

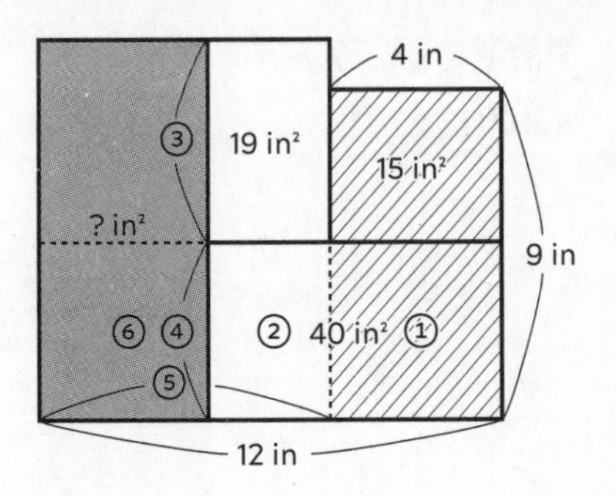

Find the area with stripes . . .
$4 \times 9 = 36$ in.^{2}

Find area ① . . . $36 - 15 = 21$ in.^{2}

Find area ② . . . $40 - 21 = 19$ in.^{2}

This is the same area as the rectangle above ②, so length ③ and length ④ are equal.

Find length ⑤ . . . $12 - 4 = 8$ in.

This is double 4 in., so the total area of ② and ⑥ must be double area ① . . . $21 \times 2 = 42$ in.^{2}

Find area ⑥ . . . $42 - 19 = 23$ in.^{2}

Area ? is double area ⑥ . . . $23 \times 2 = 46$ in.^{2}

PUZZLE 77

SOLUTION: 7 in.

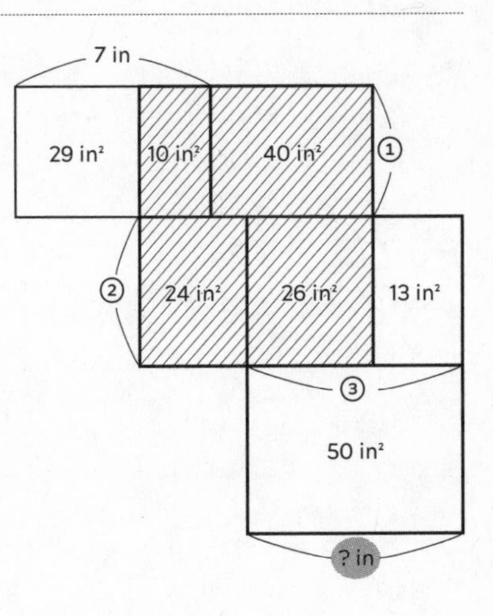

Add the top two areas with stripes . . . $10 + 40 = 50$ in.^{2}

Add the bottom two areas with stripes . . . $24 + 26 = 50$ in.^{2}

These are the same, so length ① and length ② must be equal.

Add the two areas at top left . . . $29 + 10 = 39$ in.

Add the two areas on the right in the next row . . . $26 + 13 = 39$ in.

These are the same, so length ③ must equal 7 in.

Length ? is 7 in.

PUZZLE 78

SOLUTION: 70 in.^{2}

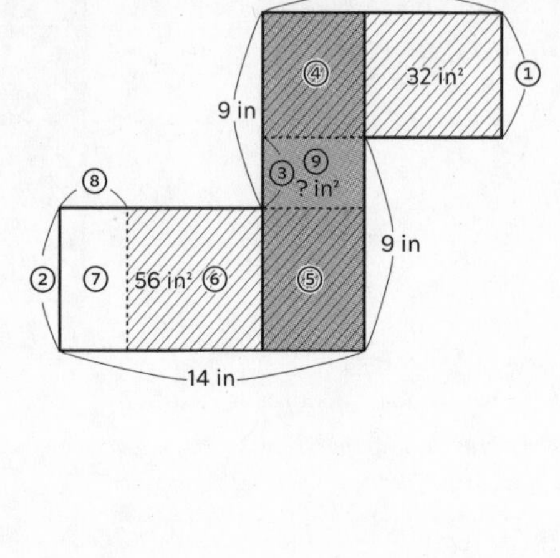

Length ① and length ② are both equal to 9 in. – ③, so they are equal to each other.

Area ④ and area ⑤ are the same height and width, so they must be identical.

Create area ⑥ of 32 in.^{2} so that the two areas with stripes are identical.

Find area ⑦ . . . 56 – 32 = 24 in.^{2}

Find length ⑧ . . . 14 – 10 = 4 in.

Find length ② . . . 24 ÷ 4 = 6 in.

Find either area with stripes . . . 6 × 10 = 60 in.^{2}

Find area ④ . . . 60 – 32 = 28 in.^{2}

Find length ③ . . . 9 – 6 = 3 in.

Because this is half of length ①, area ⑨ is half of area ④ . . . 28 ÷ 2 = 14 in.^{2}

Area ? is 28 + 14 + 28 = 70 in.^{2}

PUZZLE 79

SOLUTION: 44 in.^{2}

Find length ① . . . 9 – 7 = 2 in.

Find area ② . . . 2 × 5 = 10 in.^{2}

Find area ③ . . . 23 – 10 = 13 in.^{2}

This is the same area as the rectangle above, so length ④ is 5 in.

Find the area with stripes . . . 7 × (5 + 5) = 70 in.^{2}

Area ? is 70 – 13 – 13 = 44 in.^{2}

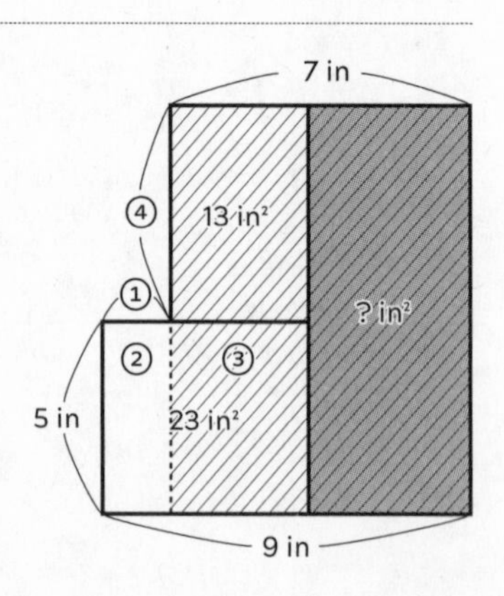

PUZZLE 80

SOLUTION: 31 in.^{2}

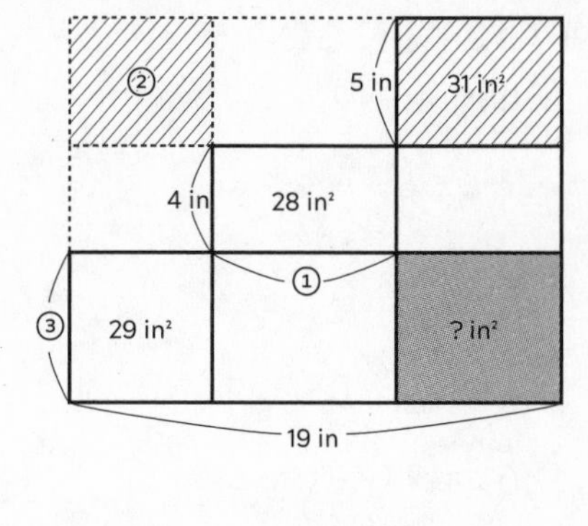

Find length ① . . . 28 ÷ 4 = 7 in.

Create area ②.

Find the combined width of the areas with stripes . . . 19 – 7 = 12 in.

Find their combined area . . . 5 × 12 = 60 in.^{2}

Find area ② . . . 60 – 31 = 29 in.^{2}

This is the same area as the bottom left rectangle, so length ③ must equal 5 in.

The top right rectangle and area ? are the same height and width, so they are identical.

Area ? is 31 in.^{2}

PUZZLE 81

SOLUTION: 50 in.^{2}

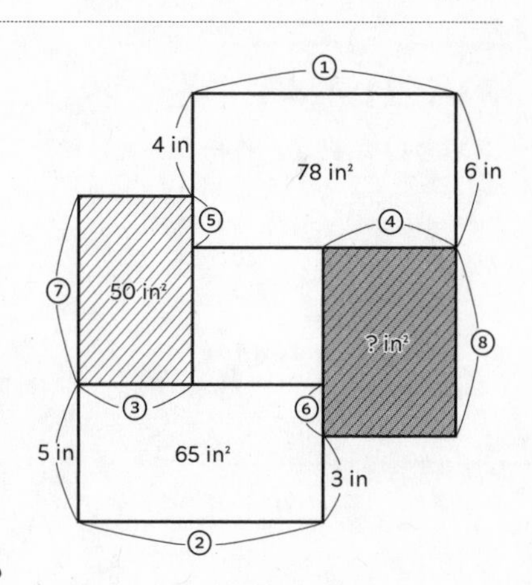

Find length ① . . . 78 ÷ 6 = 13 in.

Find length ② . . . 65 ÷ 5 = 13 in.

These are the same, so lengths ③ and ④ are also the same.

Find length ⑤ . . . 6 – 4 = 2 in.

Find length ⑥ . . . 5 – 3 = 2 in.

These are the same, so lengths ⑦ and ⑧ are also the same.

The two areas with stripes have the same height and width, so they are identical.

Area ? is 50 in.^{2}

PUZZLE 82

SOLUTION: 4 in.

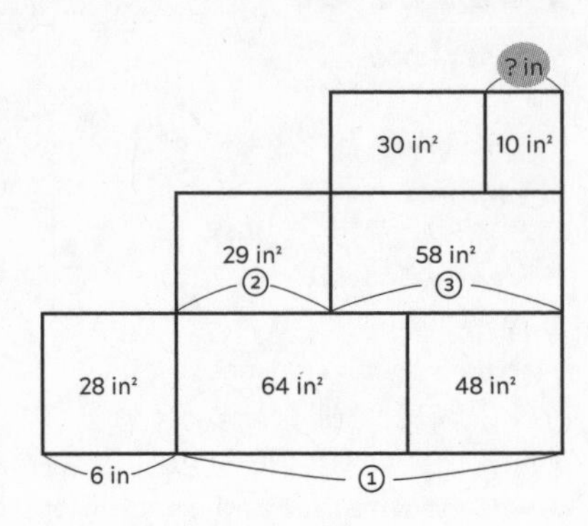

Add the two areas at bottom right . . . $64 + 48 = 112$ in.^{2}

This is four times 28 in.^{2} So, length ① is four times 6 in. . . . $6 \times 4 = 24$ in.

In the middle row, 58 in.^{2} is double 29 in.^{2} So, length ③ is double length ②, and length ① is triple length ②.

Find length ② . . . $24 \div 3 = 8$ in.

Find length ③ . . . $24 - 8 = 16$ in.

In the top row, 30 in.^{2} is triple 10 in.^{2} So, length ③ is four times length ?.

Length ? is $16 \div 4 = 4$ in.

PUZZLE 83

SOLUTION: 24 in.^{2}

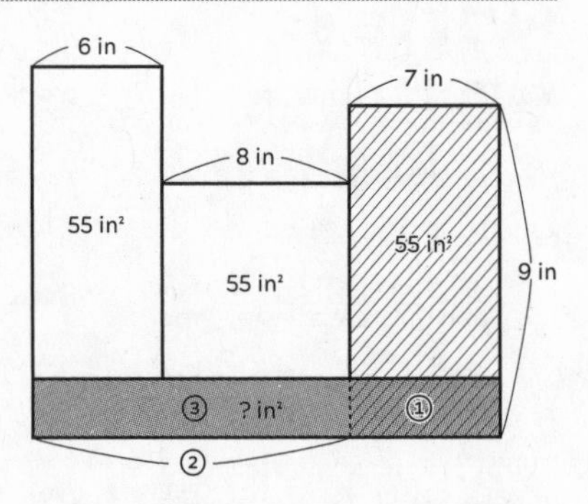

Find the area with stripes . . . $7 \times 9 = 63$ in.^{2}

Find area ① . . . $63 - 55 = 8$ in.^{2}

Find length ② . . . $6 + 8 = 14$ in.

This is double 7 in., so area ③ is double area ① . . . $8 \times 2 = 16$ in.^{2}

Area ? is $8 + 16 = 24$ in.^{2}

PUZZLE 84

SOLUTION: 7 in.

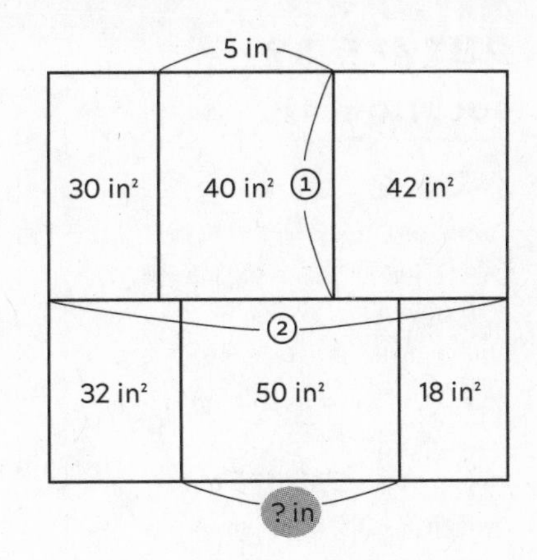

Find length ① . . . 40 ÷ 5 = 8 in.

Add the top three areas . . .
30 + 40 + 42 = 112 in.^{2}

Find length ② . . . 112 ÷ 8 = 14 in.

Add the bottom three areas . . .
32 + 50 + 18 = 100 in.^{2}

This is double 50 in.^{2} So, length ? is half of length ②.

Length ? is 14 ÷ 2 = 7 in.

PUZZLE 85

SOLUTION: 8 in.

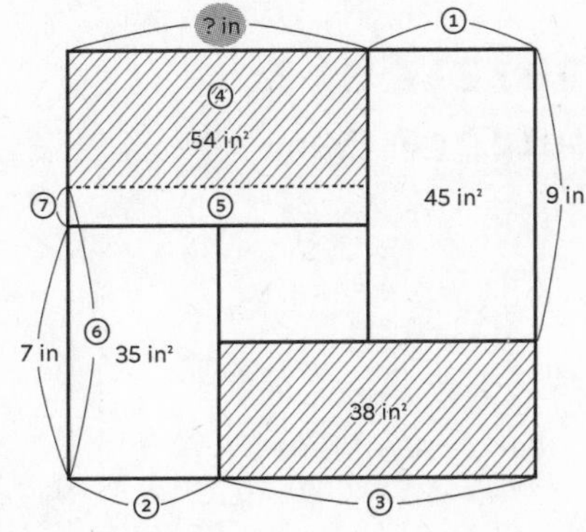

Find length ① . . . 45 ÷ 9 = 5 in.

Find length ② . . . 35 ÷ 7 = 5 in.

These are the same, so length ? and length ③ are the same.

Create area ④ of 38 in.^{2} so that the two areas with stripes are identical.

Find area ⑤ . . . 54 − 38 = 16 in.^{2}

The areas with stripes are the same height, so length ⑥ is 9 in.

Find length ⑦ . . . 9 − 7 = 2 in.

Length ? is 16 ÷ 2 = 8 in.

PUZZLE 86

SOLUTION: 4 in.

Find area ① . . . 4 × 3 = 12 in.

Find area ② . . . 25 – 12 = 13 in.^{2}

This is half of 26 in.^{2} So, length ③ is double 4 in. . . . 4 × 2 = 8 in.

Find length ④ . . . 4 + 8 – 6 = 6 in.

This is the same as the width of area ⑤, so area ⑤ is 17 in.^{2}

Find area ⑥ . . . 41 – 17 = 24 in.^{2}

Length ? is 24 ÷ 6 = 4 in.

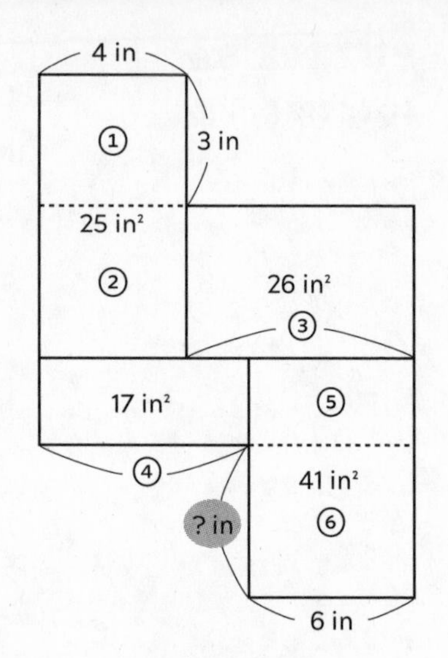

PUZZLE 87

SOLUTION: 32 in.^{2}

Create area ①.

Find the total area . . .
14 × 14 = 196 in.^{2}

Find the combined area of ① and ? . . .
196 – 11 – 33 – 64 – 17 – 17
= 54 in.^{2}

Find the area with stripes . . .
11 + 33 + 64 = 108 in.^{2}

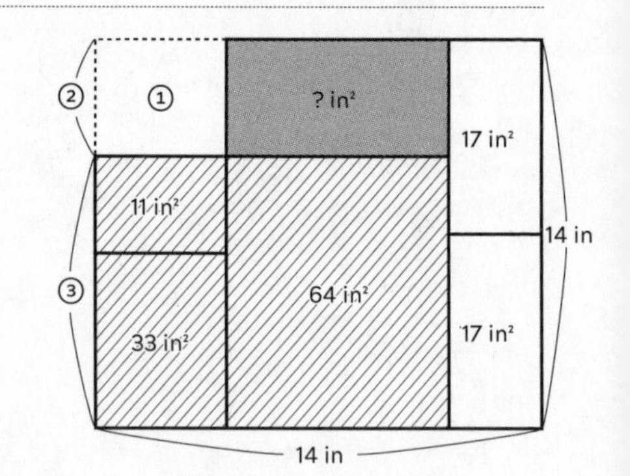

This is double 54 in.^{2} So, length ③ is double length ②.

Area ? must be half of 64 in.^{2} . . . 64 ÷ 2 = 32 in.^{2}

PUZZLE 88

SOLUTION: 13 in.

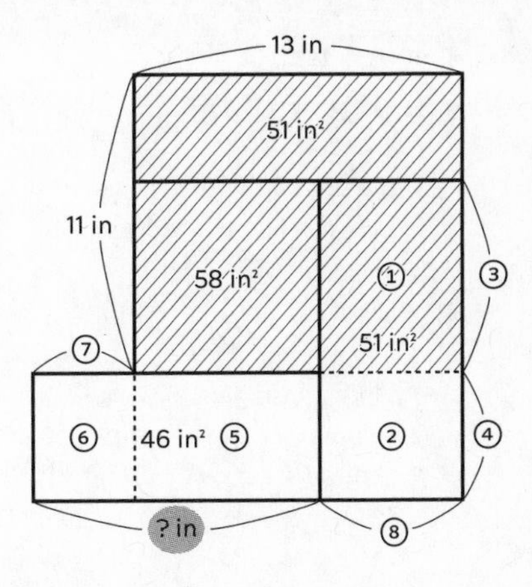

Find the area with stripes . . .
$13 \times 11 = 143$ in.^{2}

Find area ① . . .
$143 - 51 - 58 = 34$ in.^{2}

Find area ② . . . $51 - 34 = 17$ in.^{2}

This is half of area ①, so length ④ is half of length ③.

Area ⑤ must be half of 58 in.^{2} . . .
$58 \div 2 = 29$ in.^{2}

Find area ⑥ . . . $46 - 29 = 17$ in.^{2}

This is the same as area ②, so length ⑦ and length ⑧ must be equal.

Length ? is 13 in.

PUZZLE 89

SOLUTION: 3 in.

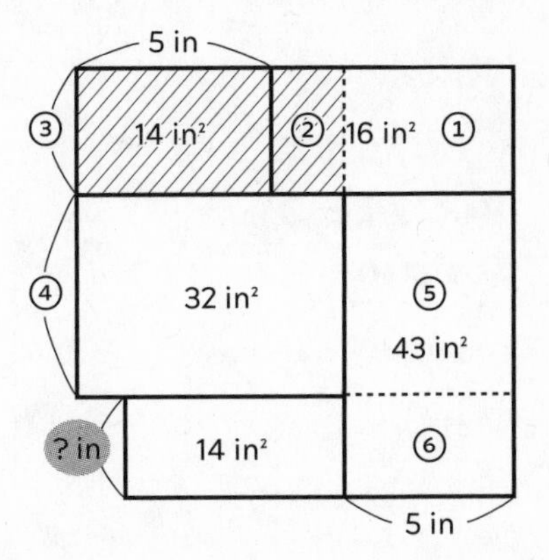

Create areas ① and ②.

Area ① is the same height and width as the top left rectangle, so it must be 14 in.^{2}

Find area ② . . . $16 - 14 = 2$ in.^{2}

Find the area with stripes . . .
$14 + 2 = 16$ in.^{2}

This is half of 32 in.^{2} So, length ④ must be double length ③.

Area ⑤ must be double area ① . . .
$14 \times 2 = 28$ in.^{2}

Find area ⑥ . . . $43 - 28 = 15$ in.^{2}

Length ? is $15 \div 5 = 3$ in.

PUZZLE 90

SOLUTION: 39 in.^{2}

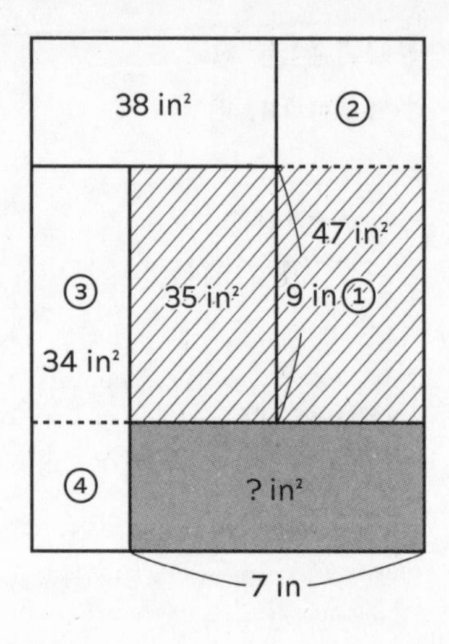

Find the area with stripes . . .
$7 \times 9 = 63$ in.^{2}

Find area ① . . . $63 - 35 = 28$ in.^{2}

Find area ② . . . $47 - 28 = 19$ in.^{2}

This is half of 38 in.^{2}, the area of the top left rectangle.

In the same way, area ① must be half the combined area of ③ and the 35 in.^{2} rectangle.

Find that combined area . . .
$28 \times 2 = 56$ in.^{2}

Find area ③ . . . $56 - 35 = 21$ in.^{2}

Find area ④ . . . $34 - 21 = 13$ in.^{2}

The area with stripes is triple area ③, so area (?) must be triple area ④.

Area (?) is $13 \times 3 = 39$ in.^{2}

PUZZLE 91

SOLUTION: 5 in.

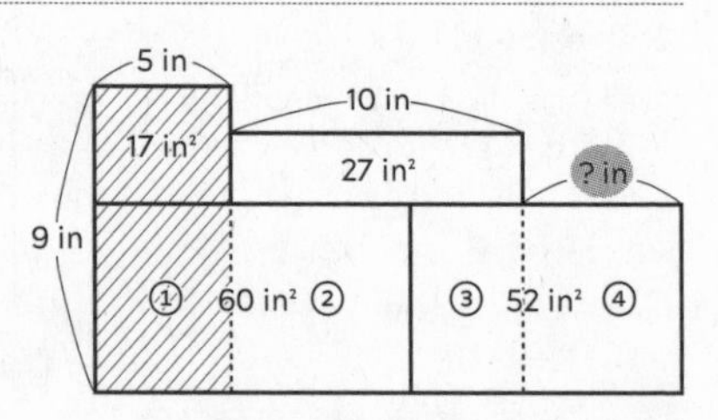

Find the area with stripes . . .
$9 \times 5 = 45$ in.^{2}

Find area ① . . . $45 - 17 = 28$ in.^{2}

Find area ② . . . $60 - 28 = 32$ in.^{2}

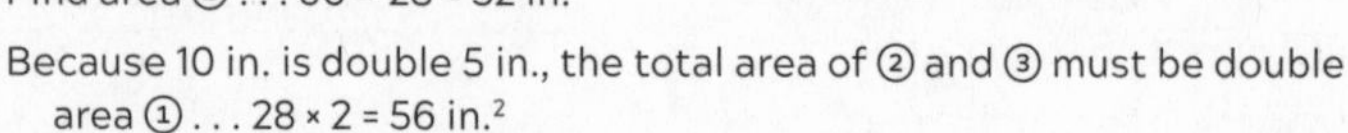

Because 10 in. is double 5 in., the total area of ② and ③ must be double area ① . . . $28 \times 2 = 56$ in.^{2}

Find area ③ . . . $56 - 32 = 24$ in.^{2}

Find area ④ . . . $52 - 24 = 28$ in.^{2}

This is the same as area ①, so length (?) must be 5 in.

PUZZLE 92

SOLUTION: 60 in.^{2}

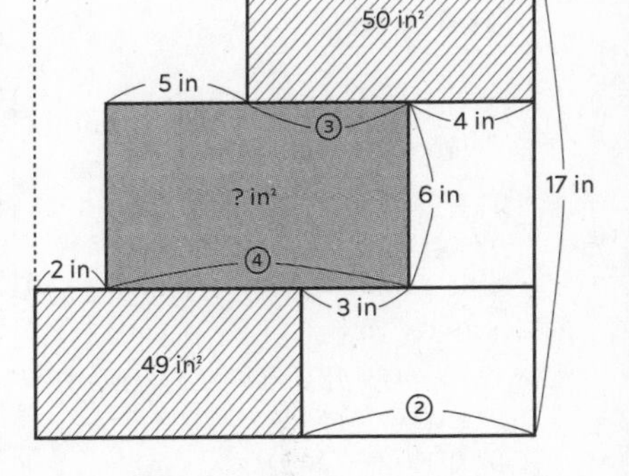

Find length ① . . . 2 + 5 = 7 in.

Find length ② . . . 3 + 4 = 7 in.

These are the same, so the two areas with stripes must be the same width.

Add the areas with stripes . . . 50 + 49 = 99 in.^{2}

Find their combined height . . . 17 − 6 = 11 in.

Find the width of either area with stripes . . . 99 ÷ 11 = 9 in.

Find length ③ . . . 9 − 4 = 5 in.

Find length ④ . . . 5 + 5 = 10 in.

Area ? is 10 × 6 = 60 in.^{2}

PUZZLE 93

SOLUTION: 29 in.^{2}

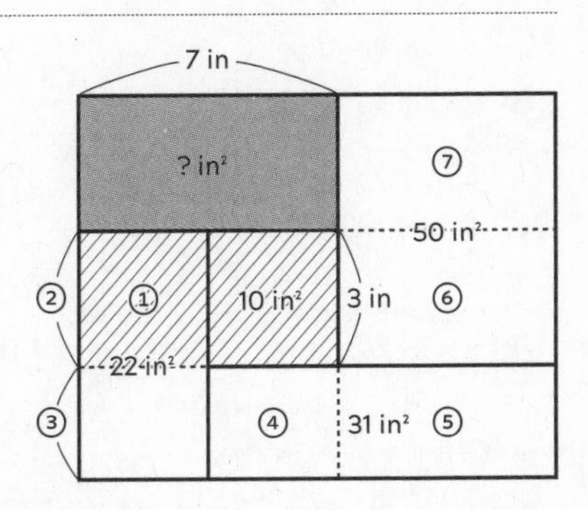

Find the area with stripes . . . 7 × 3 = 21 in.^{2}

Find area ① . . . 21 − 10 = 11 in.^{2}

This is half of 22 in.^{2} So, lengths ② and ③ are the same.

Area ④ must be 10 in.^{2}, and area ⑤ must be equal to area ⑥.

Find area ⑤ (and area ⑥) . . . 31 − 10 = 21 in.^{2}

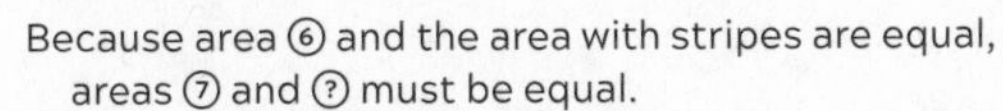

Because area ⑥ and the area with stripes are equal, areas ⑦ and ? must be equal.

Find area ⑦ . . . 50 − 21 = 29 in.^{2}

Area ? is 29 in.^{2}

PUZZLE 94

SOLUTION: 12 in.

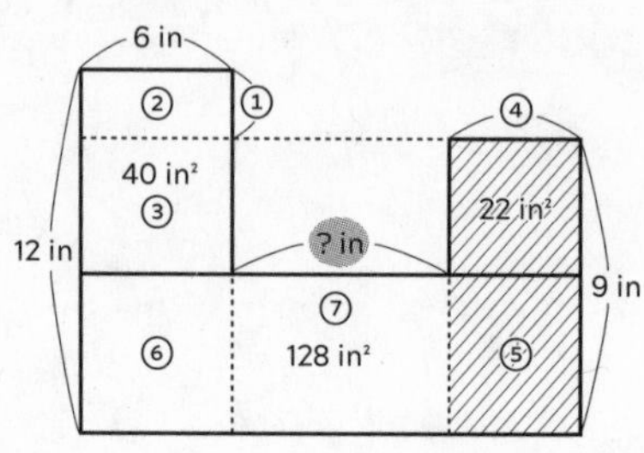

Find length ① . . . 12 – 9 = 3 in.

Find area ② . . . 6 × 3 = 18 in.^{2}

Find area ③ . . . 40 – 18 = 22 in.^{2}

This is the same as the top right area, so length ④ must be 6 in.

Find the area with stripes . . . 6 × 9 = 54 in.^{2}

Find area ⑤ . . . 54 – 22 = 32 in.^{2}

Area ⑥ is the same as area ⑤.

Find area ⑦ . . . 128 – 32 – 32 = 64 in.^{2}

This is the same as the combined areas of ⑤ and ⑥.

Length ? is 6 + 6 = 12 in.

PUZZLE 95

SOLUTION: 42 in.^{2}

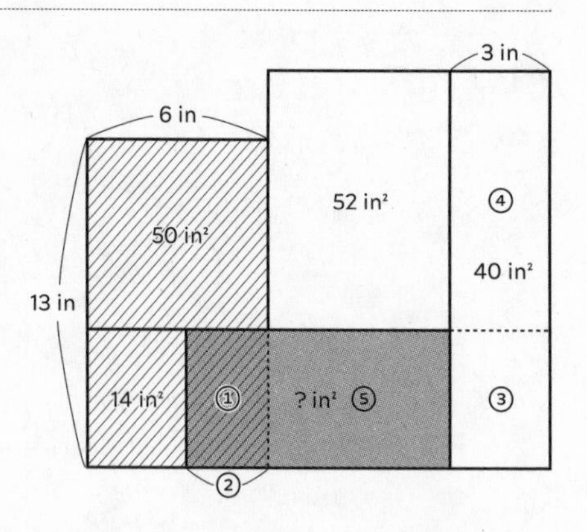

Find the area with stripes . . . 6 × 13 = 78 in.^{2}

Find area ① . . . 78 – 50 – 14 = 14 in.^{2}

Length ② must be half of 6 in. . . . 6 ÷ 2 = 3 in.

Area ③ is the same width as area ①, so it is also 14 in.^{2}

Find area ④ . . . 40 – 14 = 26 in.^{2}

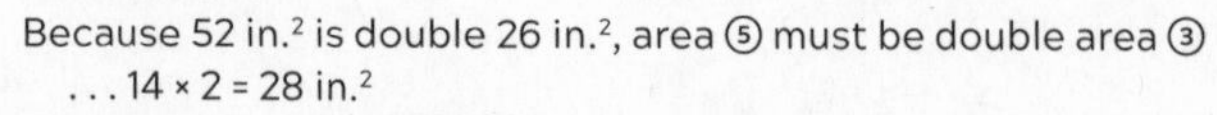

Because 52 in.^{2} is double 26 in.^{2}, area ⑤ must be double area ③ . . . 14 × 2 = 28 in.^{2}

Area ? is 14 + 28 = 42 in.^{2}

PUZZLE 96

SOLUTION: 9 in.

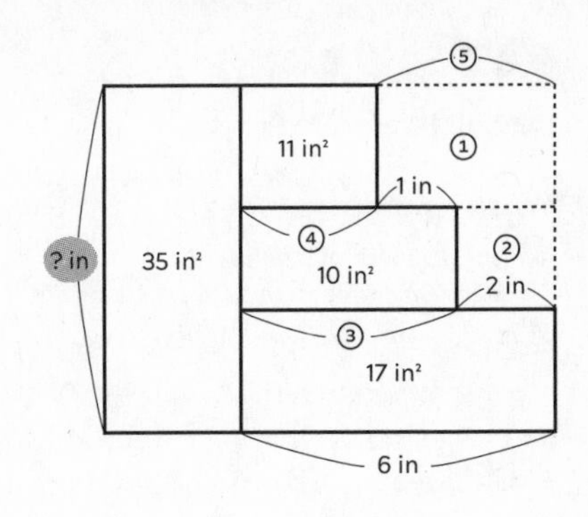

Create areas ① and ②.

Find length ③ . . . 6 – 2 = 4 in.

Find length ④ . . . 4 – 1 = 3 in.

Find length ⑤ . . . 6 – 3 = 3 in.

Area ① is the same height and width as the 11 in.² area. The total area of the top row is $11 \times 2 = 22$ in.²

Area ② is half as wide as the 10 in.² area, so it is 5 in.² The total area of the middle row is $10 + 5 = 15$ in.²

Find the total area of all three rows (ignore the 35 in.² rectangle) . . . $22 + 15 + 17 = 54$ in.²

Length ? is $54 \div 6 = 9$ in.

PUZZLE 97

SOLUTION: 80 in.²

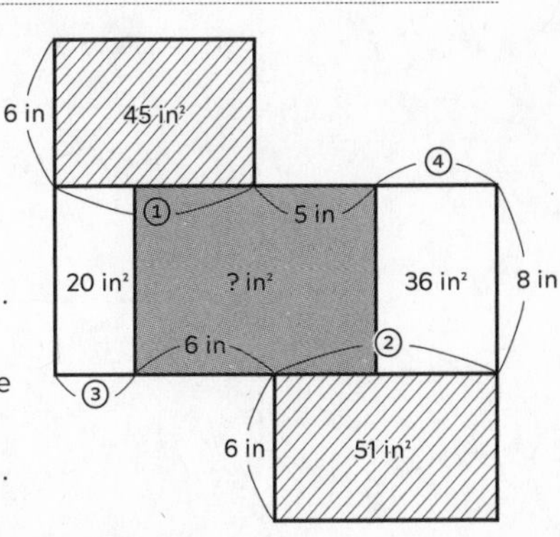

Add the two areas with stripes . . . $45 + 51 = 96$ in.²

Find their combined width (① + ②) . . . $96 \div 6 = 16$ in.

Add the two given areas in the middle row . . . $20 + 36 = 56$ in.²

Find their combined width (③ + ④) . . . $56 \div 8 = 7$ in.

The total width of the middle row is ① + 5 + ④, or ③ + 6 + ②. Add all six terms . . . (① + ②) + (③ + ④) + 5 + 6 = 16 + 7 + 11 = 34 in.

Divide by 2 to get the total width of the middle row . . . $34 \div 2 = 17$ in.

Find the total area of the middle row . . . $17 \times 8 = 136$ in.²

Area ? is $136 - 20 - 36 = 80$ in.²

PUZZLE 98

SOLUTION: 12 in.^{2}

Add the two areas on the left . . . 21 + 15 = 36 in.^{2}

This is double the area of the 18 in.^{2} rectangle, and 14 in. is double 7 in. So, length ① and length ② must be equal.

Based on this, lengths ③ and ④ must also be equal.

Add the two areas with a combined width of ③ . . . 30 + 27 = 57 in.^{2}

Add the two areas with a combined width of ④ . . . 15 + 42 = 57 in.^{2}

These are the same, so lengths ⑤ and ⑥ must be equal.

Therefore, area ⑦ is 15 in.^{2}

Find area ⑧ . . . 21 – 15 = 6 in.^{2}

The 30 in.^{2} rectangle is double area ⑦, so area ? must be double area ⑧.

Area ? is 6 × 2 = 12 in.^{2}

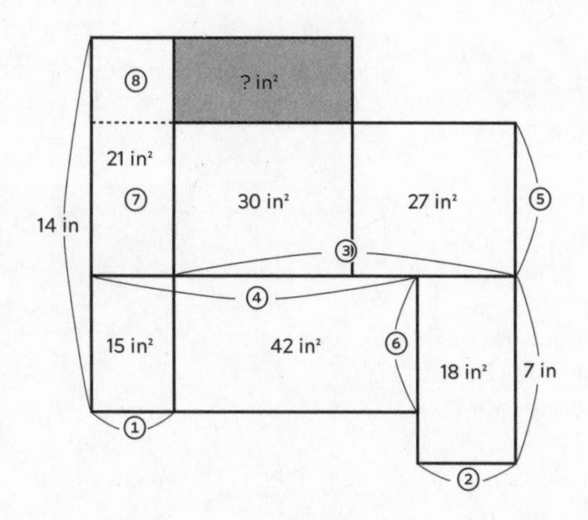

PUZZLE 99

SOLUTION: 6 in.

Imagine cutting the puzzle in half down the middle, and then placing one half on top of the other. This will create areas ① and ②, as in the figure below.

Find area ① . . . 31 – 14 = 17 $in.^2$

Find area ② . . . 37 – 20 = 17 $in.^2$

These are the same, so lengths ③ and ④ must be the same.

Find the area with stripes . . . (14 + 20) × 2 = 68 $in.^2$

Find the total area . . . (37 + 14 + 20 + 31) × 2 = 204 $in.^2$

This is triple the area with stripes.

Therefore, length ? is 18 ÷ 3 = 6 in.

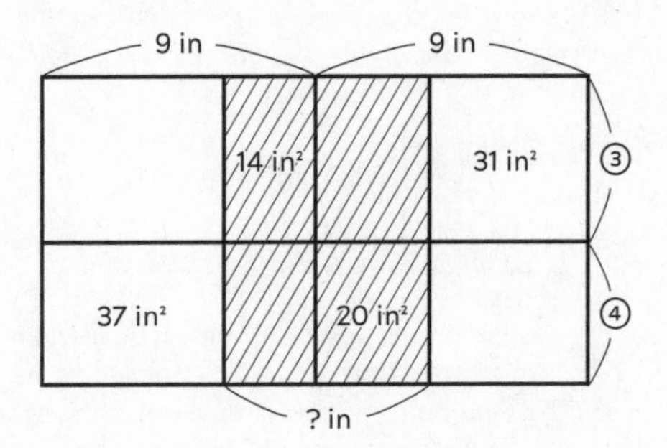

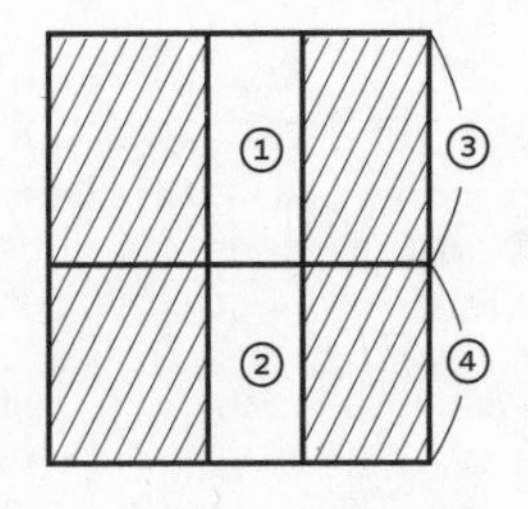

PUZZLE 100

SOLUTION: 32 in.^{2}

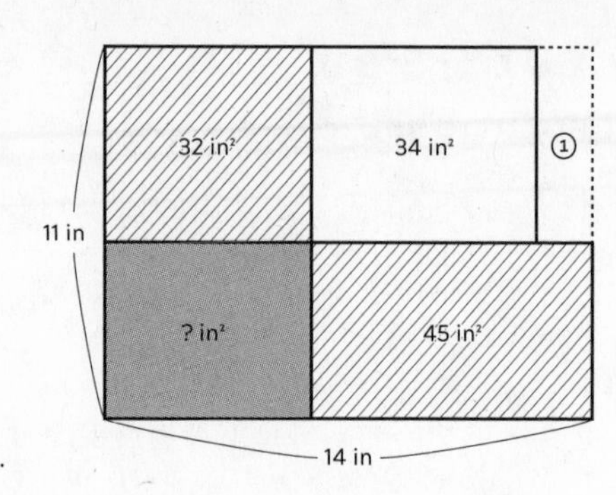

Create area ①. Consider the whole puzzle as a rectangle with four quadrants.

Find the total area . . .
$11 \times 14 = 154$ in.^{2}

Add the areas with stripes . . .
$32 + 45 = 77$ in.^{2} This is half of 154 in.^{2}

When two quadrants in diagonally opposite corners account for half of the total area, either the vertical or the horizontal boundary must divide the rectangle exactly in half. (See the proof below.)

The 34 in.^{2} area is bigger than the 32 in.^{2} area. Therefore the vertical boundary is not the one that divides the rectangle in half. It must be the horizontal boundary.

Area ? is 32 in.^{2}

PROOF

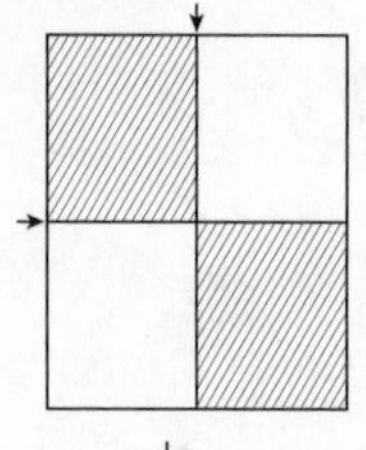

Consider a case where the vertical and horizontal boundaries both divide the rectangle exactly in half. All four quadrants will have the same area. Therefore, the area with stripes will be half the total area.

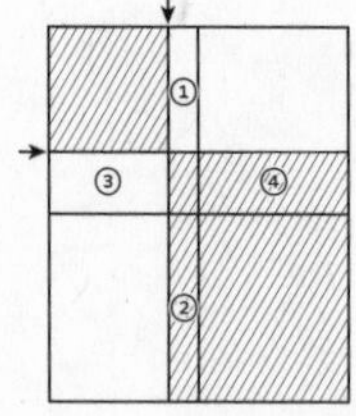

Now draw both boundaries off-center. The new area with stripes equals half the total area + ② + ④ – ① – ③. We can restate this as half the total area + (② – ①) + (④ – ③). Since area ② is bigger than area ①, and area ④ is bigger than area ③, the new area with stripes must now be more than half the total area.